D1472587

1st, 2nd, and
Next Generation
LANs

McGraw-Hill Series on Computer Communications

ISBN	AUTHOR	TITLE
0-07-003484-2	Ball	*Cost-Efficient Network Management*
0-07-005075-9	Berson	*APPC*
0-07-005076-7	Berson	*Client/Server Architecture*
0-07-005558-0	Black	*Frame Relay Networks*
0-07-005554-8	Black	*Network Management Standards*
0-07-005553-X	Black	*TCP/IP and Related Protocols*
0-07-005552-1	Black	*The V Series Recommendations*
0-07-005546-7	Black	*The X Series Recommendations*
0-07-010516-2	Cerutti	*Distributed Computing Environments*
0-07-016196-8	Dayton	*Multi-Vendor Networks*
0-07-016733-8	Dewire	*Application Development for Distributed Systems*
0-07-016732-X	Dewire	*Client/Server Computing*
0-07-019022-4	Edmunds	*SAA/LU6.2*
0-07-020346-6	Feit	*TCP/IP*
0-07-033754-3	Hebrawi	*OSI Upper Layer Standards and Practices*
0-07-028039-8	Heldman	*Future Telecommuncations*
0-07-028030-4	Heldman	*Global Telecommunications*
0-07-032385-2	Jain	*Open Systems Interconnection*
0-07-034247-4	Kessler	*ISDN, 2/e*
0-07-034243-1	Kessler	*Metropolitan Area Networks*
0-07-035134-1	Knightson	*OSI Protocol Performance Testing: IS 9646 Explained*
0-07-046455-3	Naugle	*Local Area Networking*
0-07-046461-8	Naugle	*Network Protocol Handbook*
0-07-046320-4	Nemzow	*The Ethernet Management Guide, 2/e*
0-07-046322-0	Nemzow	*FDDI Networking*
0-07-046321-2	Nemzow	*The Token-Ring Management Guide*
0-07-051104-7	Radicatti	*Electronic Mail*
0-07-054418-2	Sackett	*IBM's Token-Ring Networking Handbook*
0-07-060360-X	Spohn	*Data Network Design*
0-07-063636-2	Terplan	*Effective Management of Local Area Networks*

1st, 2nd, and Next Generation LANs

Daniel Minoli
Bell Communications Research, Inc.
and
New York University

McGraw-Hill, Inc.

New York   San Francisco   Washington, D.C.   Auckland   Bogotá
Caracas   Lisbon   London   Madrid   Mexico City   Milan
Montreal   New Delhi   San Juan   Singapore
Sydney   Tokyo   Toronto

Library of Congress Cataloging-in-Publication Data

Minoli, Daniel, date.
    1st, 2nd, and next generation LANs / Daniel Minoli.
        p.   cm.—(McGraw-Hill series on computer communications)
    Includes index.
    ISBN 0-07-042586-8
    1. Local area networks (Computer networks). I. Title. II. Title:
First, second, and next generation LANs. III. Series.
    TK5105.M556   1993
    004.6′8—dc20   93-8683
                                                      CIP

2 3 4 5 6 7 8 9 0 DOC/DOC 9 9 8 7 6 5 4

ISBN 0-07-042586-8

The sponsoring editors for this book were Neil Levine and Jeanne Glasser, the editing supervisor was Stephen M. Smith, and the production supervisor was Suzanne W. Babeuf. It was set in Century Schoolbook by McGraw-Hill's Professional Book Group composition unit.

Printed and bound by R. R. Donnelley & Sons Company.

Και ην ανηρ οικων Βρουκλυν, και ονομα αυτω Δανηλ,
και ελαβε γυναικα η ονομα Αννα, καλη σφοδρα...
After (ΣΟΥΣ) ANNA, LXX

Contents

Preface

The 1980s saw the introduction of LANs in the corporate environment for both nonproduction (back-office) and production (front-office) applications. Soon it was realized that management of the LAN required extensive tools and resources. The need to interconnect colocated and remotely located LANs has emerged as a key corporate requirement. This has given rise to completely new network technologies and services such as frame relay service (FRS), Switched Multimegabit Data Service (SMDS), and cell relay service. New applications are also pushing LAN speed up from first-generation Ethernet ranges (1–10 Mb/s), to second-generation fiber-distributed data interface (FDDI) ranges (100 Mb/s), and now to third-generation synchronous optical network/asynchronous transfer mode (SONET/ATM) ranges (Gb/s range). Many vendors see the emergence of multimedia applications as the next major turning point for LANs.

This book explores these themes from a strategic planning perspective. The field is now moving very rapidly. Practitioners cannot afford to sit comfortably on the skills they learned in the mid- or even late 1980s. We are all acquainted with "practical" LAN managers, who now find themselves struggling to learn enough open systems interconnection concepts to understand how frame relay works, given that (at least according to the trade press) many LAN interconnection networks use or will soon use this technology. These managers are just beginning to understand frame relay, when cell relay/ATM is emerging as the next-generation LAN technology; this technology will start to be deployed in 1993–1994, demanding that entirely new concepts be learned. ATM is being considered not only as the interconnection vehicle [via broadband integrated services digital network (B-ISDN)], but as the very fabric of the LAN itself.

This text can be used for a one-semester graduate course in contemporary corporate LAN studies, assuming that the students have had some previous exposure to the field, or a good treatment by the instructor. The material in this book is being used at Rutgers University and at New York University, where the author has taught for several years.

Daniel Minoli

Acknowledgments

Mr. J. A. Ladyka, Jr., president of Software Engineering Techniques, Robbinsville, New Jersey, contributed Chap. 6, Network Management Issues. Mr. G. Chanda, engineering manager for 3Com, Santa Clara, California, contributed sections of Chap. 8, which deals with FDDI. Mr. C. Owens of 3Com assisted with the review of the FDDI material. Mr. J. Lopez provided input for Secs. 1.2 and 2.1 and is hereby thanked.

The following Bellcore experts reviewed the manuscript and provided valuable comments: K. Tesink, H. Chung, and W. Beblo.

The following individuals are thanked for their moral support of this undertaking: Dick Vigilante, New York University, who also provided direct input on the Virtual College program at New York University (described in Chap. 2); Gail Allen, Rutgers University; Lance Lindstrom, DataPro Research Corporation; Tony Rizzo, *Network Computing* magazine; Joanne Dressendofer, IMEDIA; Al Tumolillo, Probe Research Corporation; and Ben Occhiogrosso, DVI Communications. Additionally, the author wishes to thank George Dobrowski, Steve M. Walters, Jim E. Holcomb, John J. Amoss, Glen H. Estes, and Tim Bouman of Bellcore.

Darren Spohn, McGraw-Hill's technical reviewer, provided excellent and thorough technical feedback on the entire book and is hereby thanked for the much-appreciated support lent to this effort. The McGraw-Hill staff is acknowledged for its assistance and much-welcome professionalism.

This book does not reflect *any* policy, position, or posture of Bell Communications Research or the Bell Operating Companies. The writing of this book was not funded by Bellcore or by the Bell Operating Companies. All ideas expressed are strictly those of the author. Data pertaining to the public switched network are based on the open literature, and were not reviewed by the Bell Operating Companies. Vendor products, services, and equipment are mentioned solely to document the *state of the art* of a given technology, and the information has not been counterverified with vendors. No material contained in this book should be construed as a recommendation of any kind.

1st, 2nd, and Next Generation LANs

Introduction

1.1 Ubiquity of Local Area Networks

The introduction of local area networks (LANs) in corporate America has advanced at a rapid pace during the past 10 years and is expected to continue to grow well into the 1990s. LAN penetration in the United States exceeded 6 million connections in 1991 and is projected to grow to 13 million connections by 1995.[1] Over 22 million personal computers (PCs) worldwide were connected to LANs in 1992.[2] The market for LANs totaled over $6.5 billion in 1991 worldwide; by 1995 the market is expected to reach $12 billion.[3] It is estimated that about 200,000 U.S. firms, out of a pool of 7 million, now have LANs and employ 79,000 LAN administrators (up from 56,000 in 1991).[2] Just in 1992, U.S. companies spent $8.9 billion for LAN support and services.

PCs and workstations now rely almost exclusively on LANs to access and disseminate information in both medium and large companies. An increasing number of small companies are also installing LANs. LANs provide a high-speed low-cost communication system over a limited distance, usually linking terminals, PCs, workstations, and resources (servers) in a single building or group of closely located buildings. Wide-area access to regional, national, and international networks is accomplished using LAN-resident communication gateways, including bridges and routers. The penetration of LANs during the past decade can be attributed to improved price and performance characteristics of both computing and local connectivity, in addition to enhanced reliability and ease of expandability. LANs provide the communications infrastructure to make the new information-intensive computing model and applications a reality. LANs also promise to support emerging multimedia and imaging technologies.

LAN technology has encompassed three generations. *First-generation* technology emerged in the mid-1970s. Many corporations are still deploying these LANs based on coaxial cable or twisted-pair cable media. *Second-generation* technology, which emerged in the late 1980s, is based on shared fiber optic cable media. *Third-generation* LANs are now becoming commercially available; they will see major deployment in the mid-1990s to support new high-bandwidth applications such as multimedia and desk-to-desk videoconferencing. The transmission speed of LANs varies from 1 to 16 Mb/s for first-generation LANs and 100 Mb/s for second-generation LANs; Gb/s rates for third-generation LANs are now under development. While data transmission is the major application to date, some LANs also carry voice and video (this is done over analog-based broadband LANs). Some digital desk-to-desk videoconferencing is also possible over traditional Ethernet LANs. Evolving cell-based gigabit LANs will enable multimedia conferencing to become a standard business interaction. (The contemporary use of the word *broadband* refers to a digital system operating at 155 Mb/s to 2.4 Gb/s; the classical use of the word *broadband* in the LAN context refers to a coaxial-based frequency-division-multiplexing communication system where one portion of the analog bandwidth is used for voice or video, and the other portion is used to carry a data signal. In this text the term has the former digital connotation.)

The proliferation of LANs has resulted in the transformation of the corporate computing environment. The 1980s witnessed the introduction of first-generation LANs for both nonproduction (back-office) and production (front-office) applications. Users who were attached to a centralized hierarchical network before 1980 are now interconnected via distributed peer-to-peer and/or client-server LAN infrastructures. Data and applications are now distributed over multiple processors. These processors communicate via reliable media such as twisted-pair cable, coaxial cable, and fiber optic cable. Wireless LANs are also being introduced. The concept of a "computing utility," where owners of PCs allow guests (in the corporation, initially) to use up their unused machine cycles, is already becoming a reality.

In turn, the need to interconnect colocated and/or remotely located LANs has emerged as a key corporate requirement of the 1990s. The trend is toward enterprisewide networking. This implies that all departments of a company are interconnected with a seamless (backbone) network, allowing unimpeded companywide access of all information and hardware resources. To connect work groups efficiently over extended enterprises, cost-effective wide-area communications services are required. This need to interconnect, coupled with the emergence of new data-intensive applications, has given rise to completely new network technologies and new services such as frame relay service (FRS) and switched multimegabit data service (SMDS). New applications are also pushing the LAN speed up from first-generation ranges (1 to 10 Mb/s for Ethernet systems and 1 to 16 Mb/s for token ring systems), to the second-generation fiber distributed data interface (FDDI) range (100 Mb/s), and now to third-generation synchronous optical network/broadband integrated services digital network (SONET/B-ISDN) range (Gb/s). Asynchronous transfer mode (ATM) is emerging as the next-generation LAN transmission technology (which is now being deployed). Many vendors see the emergence of multimedia applications as the turning point for LANs—a new generation.

Another aspect of development is global area networks (GANs). LANs have seen extension to metropolitan area network (MAN) environments. LANs also have seen national reach via wide-area network (WAN) services, including the ones just identified.* Now there is a trend toward internationally connected corporate LANs. Such GANs will become more prevalent during the next few years. By the year 2000 it is forecast that 80 percent of the computers used in business, education, and research will be linked to each other by WANs that approach the speed of today's LANs.[4]

Today a corporation can have dozens of interconnected LANs, even when these LANs are in a single building or campus. In what follows, the term *LAN* refers to the combined LAN-based enterprisewide infrastructure (the use of the singular does not imply that a company necessarily has only a single LAN).

1.1.1   What you will find in this book

This book explores the themes highlighted in the previous section from a strategic planning perspective. Table 1.1 provides a road map of the subject matter treated. Issues pertaining to common first-generation networks, such as installation, protocols, interconnection, and management, are addressed in the early chapters of the text. Second-generation FDDI LANs receive a fair coverage in the middle chapters. Evolving third-generation LANs, technologies, motivations, and approaches are given extensive coverage in the latter chapters.

The networking field, both local and remote, is now moving very rapidly. Practitioners cannot afford to sit comfortably on the skills they learned in the mid- or even late 1980s. They need to understand how frame relay works, given that many LAN interconnection net-

---

*WAN is in fact a generic term; MANs and GANs are examples of WANs.

TABLE 1.1 LAN Issues of the 1990s

⚹	**Chapter 2** The Changing Environment of the 1990s: Leading the Way to Third-Generation LANs	The LAN environment now: technology and applications; new applications
⚹	**Chapter 3** LAN Basics: First-Generation Lower Layers	MAC and LLC protocols
⚹	**Chapter 4** LAN Basics: First-Generation Upper Layers	Upper layer protocols: TCP/IP
	Chapter 5 Related LAN Standards	IEEE 802.1D; SNMP; SILS
	Chapter 6 Network Management Issues	Fault management; accounting management; configuration management; performance management; security management
	Chapter 7 Connecting Dispersed LANs	WAN services; frame relay; SMDS services
⚹	**Chapter 8** Second-Generation LANs: Fiber-distributed Data Interface	Why and when needed; fiber technology and wiring
⚹	**Chapter 9** Third-Generation LANs: Gigabit Systems	Local ATM approach; B-ISDN primer; SONET primer
	Chapter 10 Gigabit Systems for Supercomputers: Fibre Channel Standard	When to use the technology; how soon available; applications

works already or may soon use this technology, at least for the next couple of years. Managers who are just beginning to understand frame relay will have the extra challenge of learning an entirely new concept, namely ATM. ATM is being considered not only as the interconnection vehicle (via public B-ISDN services) but as the very fabric of the LAN itself. PC and workstation vendors are currently implementing multimedia capabilities in their products and expect ATM-based hubs (i.e., hubs with integrated ATM switches) to be used for local connectivity. Some feel that shared-medium technologies of traditional LANs are obsolete because of the evolving need for high throughput.

1.2 Advantages of LAN-based Computing

Stand-alone PCs, considered adequate just a few years ago, no longer fulfill the requirements of today's corporate users. Users initially found computing freedom from the centralized data processing department's mainframe and its restrictions in terms of programming languages, data structures, and limited machine cycles. However, such PC users soon found themselves isolated from real-time access to fellow coworkers and the data these coworkers generated. The stand-alone PC's data did not flow into the organization's data repository, thereby impeding strategic and synergistic utiliza-

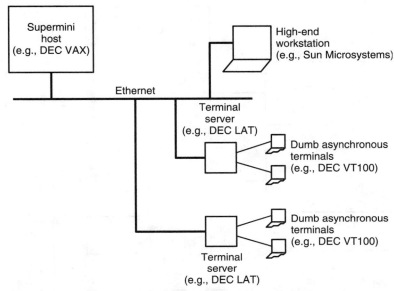

Figure 1.1 Use of LANS in minicomputer-based network (mid-1980s).

tion of such data. Consequently, in the past few years, thousands of corporations installed departmental LANs. Figure 1.1 depicts an example of a departmental LAN environment typical of the mid-1980s.

Departmental-based LAN users soon found themselves isolated once again—from users in other departments. Over time, corporations found themselves with applications running on a *variety* of platforms (from high-end mainframes, to superminis, to PCs on LANs), under the thrust of local optimization, also known as "end-user computing." In many companies, a *variety* of networks may exist, but without interworking between them. A 1990 Gartner Group study found that about 100,000 firms used three or more platforms (including LAN and non-LAN systems); it also found that about 30,000 firms had two or more distinct LANs. This predicament is now fueling the LAN-interconnection effort that is under way in many companies. Figure 1.2 depicts some of these dynamics over the years.

At first, mission-critical applications were supported mostly by dedicated IBM systems network architecture (SNA) platforms or Digital Equipment Corporation's DECnet and local area transport (LAT) structures, as seen in Fig. 1.3, right-hand side. Later, SNA started to support LANs in an ancillary mode, as shown in Fig. 1.3, left-hand side. Initially these vendor-specific platforms helped data processing mangers to (1) control applications, data, and the net-

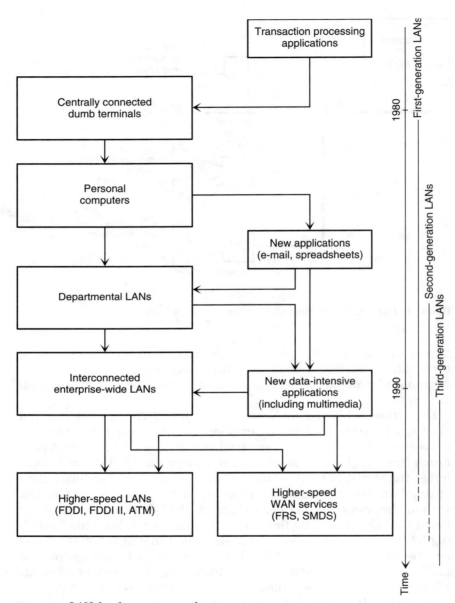

Figure 1.2   LAN developments over the years.

work; (2) provide consistent grade of service to the users; (3) ensure
reliable data integrity (security, backup, database management,
etc.); and (4) migrate to newer technologies in a consistent manner.
New business imperatives now demand openness and interoperability. Figure 1.4 shows a typical multivendor, multisystem, multiproto-

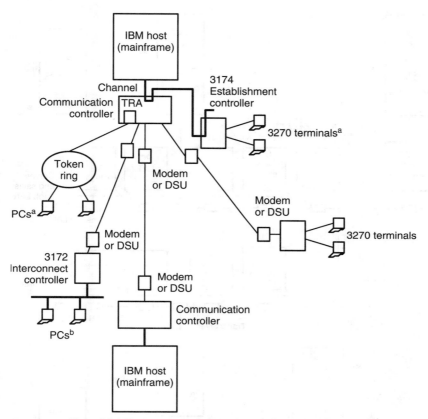

Figure 1.3 Use of LANs in mission-critical mainframe-based networks (late 1980s). TRA = token ring adapter. [The term *token ring interface controller* (TIC) was also used in the context of a 3725.] *a:* IBM's traditional approach to providing SNA mainframe access has been to require a coax board and/or 3270 terminal emulation software in each PC (except for 3270 terminals) and workstation. The devices are connected via coax or token ring to an IBM 3174 (or plug-equivalent) controller. *b:* Ethernet LAN access requires IBM 3172 interconnect controller, TCP/IP, OSI, or DECnet software on the mainframe and appropriate terminal emulation software on the PCs/workstations.

col environment of the 1990s. Internetworking departmental systems is the agenda of the 1990s. The number of installed LANs with internetwork connections is expected to rise from 164,000 LANs in 1991 to 490,000 in 1994, to 1.5 million in 1999.[5] A lingering problem of multivendor networks is the expense associated with network management; this has been one of the key reasons for relying until recently on vendor-specific platforms. A desirable goal is to have one support person for every 300 users on the network; currently, some organizations have as many as one support person for every 30 users.[6]

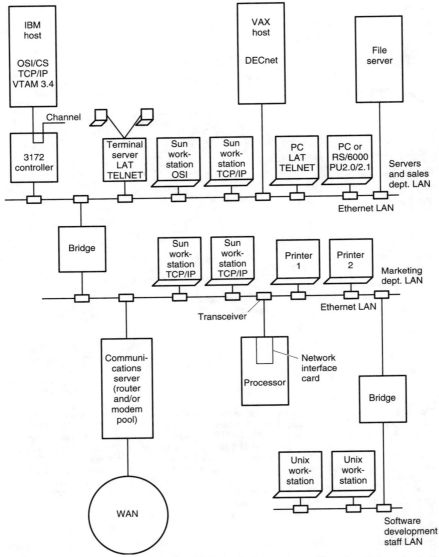

Figure 1.4 Illustrative example of integrated enterprisewide LAN. (*Note:* Network interface cards not shown in all devices.)

The organizational advantages of migrating to a LAN-based PC-workstation environment can be grouped into three categories:

1. Strategic business advantages
2. Tactical and/or operational advantages
3. End-user advantages

1.2.1 Strategic business advantages

There are corporate-level advantages in LAN-based computing. Some of these include:

1. *Facilitating corporate communication.*   Underlying the effective functioning of any organization is proper communication. Information exchange is facilitated with LANs, rather than stand-alone PCs. A LAN allows a corporate manager access to an up-to-date integrated repository of information that may be needed for decision making. Information exchange can also be accomplished with stand-alone PCs by gathering the needed information from the individual PCs. This gathering process is, at best, time-consuming; it is no longer the practice in most companies, particularly as businesses seek to improve white-collar productivity. In addition, coworker communication is expanding from just exchanging textual information to sending video, image, and multimedia; LANs can support such expanded communication.

2. *Improving competitiveness.*   Time-to-market, product quality, reduced manufacturing and production cost, rapid product adaptation, intensive data and market analysis, increased internal and external accountability, and improved customer service are all factors and measures of how well a company can meet the increased competitive climate of the 1990s. Expeditious and unimpeded access to corporate data, facilitated by LANs, is a key ingredient of this competitive drive, and one which lowers costs and increases revenues. *Computing power* is a term that has become common in today's business world.

3. *Improving work group dynamics.*   The use of LANs fosters organizational synergy. LANs can enhance the way organizations research, develop, market, and sell their products. The atmosphere of cooperative processing within a work group supported by a LAN, through what is sometimes called *groupware,* allows an employee to contribute effectively toward the completion of an organization's undertaking. The consolidation of input from a number of team members produces a result that is usually superior in quality to that which can be produced by a single employee. The task can also be completed in less time. This allows companies to share resources, including expensive human resources, and/or utilize them more efficiently.

4. *Reducing data processing budgets.*   Organizations are constantly seeking ways of reducing expenses and increasing productivity. LANs can support these organizational goals. As a result of competition within the LAN industry, software and hardware prices are being driven down. LAN products are flexible, allowing the organization to improve networking capabilities without sacrificing existing hardware and software investments. As prices decrease and capabili-

ties increase, organizations realize that they can off-load tasks onto LAN-based PCs, thereby saving mainframe machine cycles, and, in turn, cap mainframe expansion costs. LANs allow the connection of multivendor products. Studies have shown that a task can be accomplished through end-user computing for less money. However, when making cost comparisons, costs embedded within mainframe operation must be appropriately evaluated. For example, mainframe operations costs include network management, data backup, disaster recovery, physical security, and so on—costs which are not always included in the end-user computing economics calculations.

1.2.2 Tactical and operational advantages

There are management-level advantages in LAN-based computing. Some of these include:

1. *Reducing per-user computing costs.*   The sharing of peripherals, software, and centralized storage results in cost savings. Stand-alone PCs do not share peripherals, software, or hard-disk storage with other users. Peripherals, such as high-quality color laser printers (600 to 1200 dots per inch), projection screens, and so on are expensive. It is difficult to justify the cost of these peripherals with a single user. Printers, modem pools, communication gateways, and links are not constantly in use; therefore sharing these peripherals makes sense. As the capacity of a file server increases, the cost decreases because of efficiencies in packing (power supply, control unit, console, etc.). Deploying a large, centralized server can decrease the amount of local storage needed at each user's location, although, in some environments, users continue to prefer utilization of local storage. (Some local PC-workstation storage, however, is necessary since the operating system needs memory to support paging. Network performance is currently not adequate to support network-based paging.) Many vendors offer "network versions" of software applications which are shared by the various users. The per-user cost of network-based applications is lower than the cost of individual purchases of the application across a work group.

2. *Improving computing redundancy.*   Increasingly, key LAN devices have a "hot spare" directly attached to the network. For example, if the file server fails, some sophisticated networks have a backup file server that is automatically activated with mirrored information. In other cases, the main server's drive is switched over to the backup at the time of service failure.

3. *Improving software management.*   Having to update applications on individual stand-alone PCs when new versions of software

become available is time-consuming in a production environment. In a LAN-based environment, updating applications is done through a central entry point. Also, organizations with multiple stand-alone PCs may find themselves with different versions of the same software application; multiple versions may lead to compatibility and support problems. In a client-server architecture, applications can be allocated among logical components. Presentation services, user interfaces, and data can be placed on either the server or the client's device depending on organizational needs (proper partitioning of applications results in reduced LAN traffic).

4. *Improving data integrity.*   Users with stand-alone PCs do not always have the time to back up their data on a routine basis. A server can be programmed to automatically back up the data that is being transferred from the client to the server. The server can also back up the data on the individual client's hard disk. Many databases allow users to share data by creating a locking feature on their application. Locking features prevent a user from updating a database record while another user is writing to the same record. This feature provides consistent data for the users and the organization. Additionally, database servers contain controls to ensure that entered data conforms to preset parameters.

5. *Improving response time.*   A properly configured LAN can improve response time by allowing resources on other processors to be made available to the instantaneous needs of a given user. While the technology to support distributed computing is just beginning to appear, the response time is also improved in basic LAN arrangements. This is because the mainframe is off-loaded, taking its operation away from the "knee" of the delay-load curve, and because PCs tend to provide the users with more consistent performance. Naturally, the number of users connected to a single LAN segment must be kept to a reasonable level. When it becomes congested, the LAN can be partitioned into sub-LANs connected with bridges or routers.

1.2.3   End-user advantages

There are user-level advantages in LAN-based computing. Some of these include:

1. *Enhancing computing flexibility.*   Many of today's LANs are based on portability of information across dissimilar platforms. This open and flexible architecture allows the coexistence of platforms whereby users enjoy the flexibility of selecting an operating system for their PCs that best suits their particular needs.

2. *Simplifying use.*   Ease of use allows the users to complete their work faster and more efficiently. Many LAN-based database applications use easy-to-follow menu and graphical interfaces. If needed, the front-end application controlling the user interface can be changed while the back-end database remains the same. This allows the user or a software developer to customize the screen to simplify user access. Conversely, the database application can be changed, while keeping a familiar user interface.

3. *Augmenting the application repertoire.*   On a stand-alone PC users need to purchase their own software. In practical terms, this means access to a handful of applications and the management of each application individually. With a LAN-based configuration, users have access to a broad range of applications residing on a server that can enhance and complement their working environment.

4. *Expanding reporting capabilities.*   The user's ability to create a variety of reports is enhanced when there is access to LAN-based resources. Access to the corporate data repository enables users to create comprehensive reports with reliable and consistent data. These reports can be generated using central server software supporting reporting applications.

5. *Enhancing security.*   Sensitive user data can be protected by the security facilities included in the network operating system (NOS). Among other features, NOSs restrict log-ons to specific nodes and enforce periodic password changes. Security is also provided by the use of diskless workstations. These workstations rely exclusively on LAN-based servers for programs and data, thereby eliminating the risk of unauthorized access and dissemination of information.

6. *Greater access to training and learning.*   The LAN support staff of an organization can provide on-line training programs that are user-specific. These programs provide easy-to-follow training databases supporting a range of topics for the users.

The advantages identified above will continue to drive deployment of LANs in the corporate environment for the foreseeable future.

1.3   Keeping Up with the Times

LAN-based computing is a powerful tool to control corporate costs, enhance productivity, and reduce time-to-market delays. However, this technology is in constant motion. The LAN manager needs to understand where this field is going and when it will reach certain milestones over the next two to four years. The purpose of this book is to enable the LAN manager to acquire such perspective. This process starts with the next chapter, which provides an overview and motivation for some of the topics which are discussed in later chapters in detail. It also sets the common basis of knowledge required for an effective reading of the balance of the material. For a more extensive description of the environment of the 1990s, particularly enter-prisewide networking, the reader may refer to recent treatises.[7,8]

References

1. *The U.S. Market for LAN Interconnect,* Frost & Sullivan, New York, 1991.
2. J. Mulqueen, "LAN Experts Are in Big Demand," *Communications Week,* March 9, 1992, page 5.
3. *Local Area Newsletter,* Information Gatekeepers, Boston, Mass., June 1991.
4. D. Hartman, "Unclogging Distributed Computing," IEEE Spectrum, May 1992, pp. 36 ff.
5. J. S. Skorupa, "Evaluation of Intelligent Wiring," COMNET 92, Washington, D.C., January 1992.
6. *Proceedings of the NOMS 92 Conference,* Minneapolis, IEEE, 1992.
7. D. Minoli, *Telecommunications Technology Handbook,* Artech House, Norwood, Mass., 1991.
8. D. Minoli, *Enterprise Networking, Fractional T1 to SONET, Frame Relay to B-ISDN,* Artech House, Norwood, Mass., 1993.

2

The Changing Environment of the 1990s: Leading the Way to Third-Generation LANs

This chapter describes the LAN environment as it has evolved in the recent past. In doing so, it highlights important areas of current movement and transition within the corporate networking realm.

2.1 The PC-LAN Environment

2.1.1 Three generations of LANs

First-generation LANs. First-generation LANs were developed in the early 1970s to provide what was then considered high-speed local connectivity among user devices. A distributed medium-sharing discipline, imported from studies in packet radio transmission, has been employed. The Ethernet technology was brought to the market by a joint effort among Xerox, Intel, and Digital Equipment Corporation. Ethernet was designed in 1973 at Xerox's Palo Alto Research Center as a network for minicomputers. The name *Ethernet* derives from the (incorrect) nineteenth-century theory that electromagnetic energy is transmitted through a fluid substance that permeates the universe, called the ether. Ethernet products reached the marketplace in the early 1980s. Ethernet initially employed coaxial cable arranged in a logical bus, operating at 10 Mb/s. Now, thin coaxial and twisted-pair cable can also be used. Ethernet systems have been widely employed in engineering, research, and manufacturing environments.

At first, there were no common standards for LANs, and different companies utilized different approaches. Extensive standardization work has been done by the Institute of Electrical and Electronic Engineers (IEEE) in the past 15 years, leading to well-known standards such as the IEEE 802.2, 802.3, 802.4, and 802.5 LAN standards (discussed in Chap. 3). Thin coax and twisted-pair media have also been standardized (i.e., specifications are available), while the use of fiber cable has not been standardized. In the early 1980s, a token bus and a token ring technology were also standardized, operating at 1 Mb/s and 4 Mb/s. The token medium-sharing discipline is a variant of the polling method common in traditional data networks. Just as in a polled network, only the LAN user possessing a special packet, the token, can transmit; however, instead of centrally controlled polling, the token is passed from station to station in a fairly equitable manner. Token-based LANs saw penetration in office environments for access to IBM's mainframes. Token ring systems took the approach of using (shielded) twisted-pair wires as the underlying medium, mainly because such a medium is cheaper and simpler to install than coaxial cable.

Over the decade, the cost of connecting a user to a LAN decreased from about $1000 to less than $200. Ethernet cards costing $100 are appearing on the market.[1] After a relatively slow start in the late 1980s, 16-Mb/s token ring adapter cards have now become a commodity item; they range in price from $700 to $900.[2]

Unshielded twisted-pair (UTP) is now the dominant LAN medium for traditional LANs. In 1990, shielded twisted-pair (STP) was in place in 30 percent of all LAN installations worldwide; UTP was in place in 24 percent of all LANs; the balance included thin and thick baseband coaxial, broadband coaxial, fiber optic cable, and wireless. By 1995, UTP is forecast to be in place in 37 percent of the installations, STP in 21 percent of the installations, and other media (notably, fiber optic cable and wireless) in 42 percent of the installations.[3]

There have been various degrees of LAN-generated synergy in the corporate environment. In a *low-enhancement environment,* LANs are used principally to send data to printers and for unstructured intracompany messaging. When communicating with a host, the PC either is "gatewayed" away from the LAN and into the host's own network, or is required to emulate a dumb terminal. Peer-to-peer architectures (Sec. 2.1.7) support a *medium-enhancement* environment, where the PC can act as a peer to the mainframe; the PC can support the entire communication protocol stack, including, for example, supporting file transfer, remote database access, transaction processing, and so on. The client-server architectures (Sec. 2.1.7) support a *high-enhancement* environment allowing truly distributed and cooperative computing.

Second-generation LANs. Higher network performance is required in order to support the applications now being put on line by the users.

Throughput is a key requirement. Traffic generated by LAN-based applications ranges from a few kilobytes, as in e-mail applications, to several megabytes, as with high-resolution graphics applications. One way of increasing the bandwidth available to applications is to replace the existing network with one based on FDDI. Efforts on *second-generation* LANs started in the early 1980s; products began to enter the market in the late 1980s. This token-based backbone-campus technology extended LANs' features in terms of the geographic radius, now covering a campus, as well as in the speed, now reaching 100 Mb/s. Implementors eventually settled on multimode fiber as the underlying medium, although support for single-mode fiber was added in the late 1980s. The interface card costs have slowed down the deployment of FDDI systems. The cost of connecting a user to a FDDI LAN started out at about $8000 and is now around $1500 to $2000. Some forecast line card costs of $800 to $1000 by 1994.[4] FDDI concentrators are typically employed to keep the cost down. Efforts to facilitate the use of twisted-pair copper wires for FDDI are now under way, in order to bring the station access cost down (copper-based interfaces cost several hundred dollars). While standards work in this arena has been slow in picking up speed, vendor-based products are becoming available. For example, in mid-1992, IBM announced shielded-pair distributed data interface (SDDI); other companies also provide similar technology.[4] In June 1992, after much discussion, the X3T9.5 subcommittee of ANSI endorsed a method proposed by Crescendo Communications Inc. [specifically, multiline transmission 3 (MLT-3)] for data-grade unshielded twisted-pair (DTP) transmission; final acceptance of the standard was expected by 1993. At publication time there were also suggestions for a new 100 Mb/s Ethernet technology.

Efforts were under way in the mid- to late 1980s to standardize a MAN technology extending LAN capabilities to metropolitan environments. This culminated in 1990 with the acceptance of the IEEE 802.6 standards based on distributed queue dual bus (DQDB) principles. DQDB has a similar cell structure as ATM and is used, for example, in SMDS access links. IEEE 802.6 only receives limited treatment in this book.

Third-generation LANs. Starting in 1990, efforts have been under way to develop *third-generation* LANs supporting gigabit-per-second speeds over fiber facilities. Two efforts are under way. The first is based on extensions to FDDI; the project is known as FDDI follow-on LAN (FFOL) and is sponsored by ANSI. The second, more recent, and perhaps more successful effort is based on ATM principles; the project is known as local ATM (LATM) and is sponsored by industry vendors. Table 2.1 depicts some trade press headlines on ATM technology at writing time. According to some proponents, 70 percent of all networked

TABLE 2.1 ATM Headlines at Writing Time

Communications Week, May 11, 1992	"Adaptive Targets Local ATM with a Switch Called Wanda"
Communications Week, May 18, 1992	"Key Interoperability Spec Completed— ATM Forum Clears Path for Device-to-Switch Interoperability"
	"The Dawn of ATM Networking"
	"ATM Is Coming—Be Ready"
	"Digital Link Readies DSU for ATM Networks"
	"ATM Products Span the Enterprise Network"
Communications Week, May 25, 1992	"Cell Relay Service Planned"
Communications Week, June 8, 1992	"Cabletron Working on ATM Switch and Interfaces"
	"New Fibermux Hub to Bolster Backbone Speeds with ATM"
Communications Week, June 15, 1992	"GTE Government Unit Has ATM Switch"
	"BBN and U-B Tie ATM Knot"
	"ATM Promises Higher Speeds"
Communications Week, June 22, 1992	"ATM Switch Has Flexible Architecture"
Communications Week, June 29, 1992	"AT&T Outlines Switch Plans"
	"New Gear...Vendor Offering Include ATM Switches"
	"Adaptive Ready with ATM Switch"
	"Consortium Starts Gigabit-Net Tests"
Communications Week January 4, 1993	"Cabletron Readies ATM Card for Delivery"
Communications Week, February 1, 1993	"ATM Switch Pact"
Communications Week, February 8, 1993	"Newbridge Plans ATM Hub Blitz"
	"Users: Private ATM to Come First"
	"GDC, Netcomm Partner on ATM"
Communications Week, February 15, 1993	"What's the Story on ATM?"
Communications Week, February 22, 1993	"A Cascade of ATM"
	"Fibermux Plots ATM Course"
	"IBM IN to Handle Voice via ATM"
	"Users Seek Details on ATM Services"
Communications Week, March 1, 1993	"Newbridge Unveils ATM Family for LAN"
	"ATM Contract Up for Grabs"
Communications Week, March 8, 1993	"IBM Pitches Low-Speed ATM"
	Sun and SynOptics Pair to Develop ATM Adapters and More"
Communications Week, March 15, 1993	"Low Cost ATM Product Unveiled"
Communications Week, March 22, 1993	"IBM Keeps Pushing Low-Speed ATM"
	"Hughes to Develop ATM Matrix Switch for Enterprise"
Communications Week, March 29, 1993	"AT&T & HP Propose 50-Mbps ATM"
Communications Week, April 5, 1993	"UTP Working Group Studies Low-Speed ATM Proposals"

NOTE: Although the information in this table is time-dependent, it illustrates the early high level of interest in ATM technology and services from a variety of suppliers.

workstations will use ATM technology in the year 2000. ATM switches to support high-end workstations were already appearing in mid-1992. Several vendors, including Adaptive, Ascom Timeplex, BBN Communications, Newbridge Communication, and Stratacom, were targeting products by mid-1993.[5] Workstation manufacturers (including NeXT and Sun Microsystems) were reported to be developing interface cards to connect their equipment to ATM switches. Initial costs may be around $4000 per port, but these should come down considerably in the next couple of years as chipsets emerge. Already, in 1992, some carriers were planning wide-area ATM networks, and some users were installing or planning to install products based on this technology.[5,6]

There have been recent demonstrations showing the feasibility of transmitting 155 Mb/s SONET/ATM signal over 131 meters of unshielded 24 American Wire Gauge (AWG) single twisted-pair copper cable and over 213 meters of shielded 22 AWG twisted-pair cable with a bit error rate (BER) no greater than 10^{-13}. Issues of meeting FCC radiation specifications are under study.[7] Some investigators are also looking to see if the recently proposed MLT-3 FDDI twisted-pair method can be brought up to 155 Mb/s; there is enough signal margin to make this possible in principle.

One should not assume, however, that traditional LANs will disappear from the business landscape, anymore than low-speed analog private lines disappeared from private networks when high-speed digital lines became available, particularly for the access portion of the network (after all, a banking terminal or a lottery ticket terminal does not need a 45 Mb/s data rate—they actually get by with less than 100 b/s). There will be a continued need for text-based business functions (as discussed in Sec. 2.3.1). However, some companies are moving to image-based operations (for example, claims processing with digitized photos of casualty and damage, check processing, signature verification, etc.), requiring more bandwidth (see Sec. 2.3.2).

Third-generation LANs are discussed in more detail in Chaps. 9 and 10. Table 2.2 summarizes some of the features of these three LAN generations.

Wireless LANs. Wireless LANs allow users to send and receive data within a building without having to install a cabling plant. Another advantage is the ability to easily add users as needed. Transmission technologies include spread-spectrum radio, microcell approaches, and directional and diffused infrared (signal reflection off walls, ceiling, and floor). The medium-sharing principles embodied in Ethernet LANs were initially developed to support digital (military) radio communications (e.g., the Aloha system); it is interesting that the technology is returning to its roots.

Typical wireless-LAN applications include data collection for ware-

TABLE 2.2 Typical features of LANs

Generation	Speed (Mb/s)	Equipment	Interconnection speed/services	Applications
First	1–16 (Ethernet token ring)	Terminals; PCs; workstations	9.6 kb/s; 56 kb/s; T1; frame relay SMDS	Office automation; decision support for business functions as accounting (spreadsheets), project management, etc., mainframe access; manufacturing; some graphical applications
Second	50–100 (Arcnet FDDI)	PCs; high-end workstations; high-end servers (image servers)	Fractional T1; T1; T3; SMDS	Backbone interconnection of LANs; CAD/CAM graphics; imaging
Third	150–1200	High-end workstations; video equipment; high-end servers	SONET; B-ISDN/cell relay SMDS	Multimedia; desk-to-desk multimedia conferencing; multimedia messaging; CAD/CAM; visualization; animation; medical imaging; digitalvideo interactive LAN-based training; supercomputer and scientific applications (such as meteorological, radar, etc.)

house inventory and stock exchange trading floor data transmission. Wireless LANs are still slower than wired Ethernet LANs: 2 to 5 Mb/s at the high end (at press time), slower for some products. Wireless LANs also support smaller groups compared to cabled LANs. Also they are more expensive, ranging from $500 to $900 per user; however, proponents see these costs coming down.[8] Typical transmission ranges are 90 to 250 m (300 to 800 feet). Usually the radio transceiver provides a link to a set of backbone hubs. Vendors include BICC Communications, Photonics Corp., and Windata Inc. While early applications handled mostly data, proponents see a uniform mixture of voice and data by 1997. The wireless LAN market is put at $300 million in 1993 by some, $150 million by others.[9,10]

Wireless LANs now support greater speed than just a few years ago and are becoming more compatible with traditional Ethernet or token ring LANs. Some wireless LANs can be interconnected with Ethernet LANs in the same or different buildings. Developments in the next couple of years will include speed consistent with traditional LANs (up to 16 Mb/s, perhaps even 100 Mb/s)[11]; integrated network management; greater user mobility; access to any peripheral whether on the wireless segment of an interconnected LAN or on the cabled side. The IEEE 802.11 committee on wireless data networks is now working on developing a standard. Resolution, however, is not imminent.

2.1.2 PCs

LAN-based applications increasingly involve the exchange of relatively large files. The ability to move and process large amounts of data in support of new user applications can be traced to developments in microcomputing hardware in the users' PCs and workstations, and in operating system advances. In particular, improved microprocessors support operating systems and applications that address more memory without having to rely heavily on "slow" memory pagination methods.

In turn, the requirement to move large files raises the issue of *nominal* and, more importantly, *effective throughput* on the LAN. Nominal throughput is dictated by the physical LAN medium and the clocking mechanism operating at the physical level. Effective throughput depends on the protocols used, not only at the medium access-control (MAC) level to control collisions for medium sharing but also at the higher layers. Some studies have indicated that the effective LAN throughput on traditional LANs using off-the-shelf hardware can be as low as between 35 percent and 60 percent of the nominal throughput. The issue of throughput becomes even more pronounced when two geographically separated LANs need to be connected by a wide-area facility. Whereas a couple of years ago, connection speeds of 56 kb/s were adequate, new applications are driving the connection speed requirements up to 1.544 Mb/s, 45 Mb/s, and even higher.

The following sections survey some of the characteristics of users' devices typically connected to a LAN, namely PCs and workstations. This is followed by a discussion of other support hardware (for example, servers). Software on the PCs and supporting the LAN is discussed next.

PCs and workstations incorporate a very large scale integration (VLSI) microprocessor and supporting subsystems. Microprocessors (also, commonly, but incorrectly, known as central processing units, or CPUs) are the most critical component of PCs in terms of determining the power of the computing platform. The processor's major functions are to execute program instructions, read and write information to memory, and access peripheral devices.

Key U.S. manufacturers of microprocessors include Intel and Motorola. Intel's microprocessors include the 80386SX, 80386DX, 80386SL, 80486SX, and 80486DX models. Intel's microprocessors are used in IBM PCs and compatibles. Motorola's microprocessors include the 68000, 68020, 68030, and 68040 models. Motorola's microprocessors are used in Macintosh computers and high-end workstations. At the time of writing, 80486SX chips sold for $120 in quantities of 1000 or more.

The performance of a microprocessor can be measured by (1) the clock speed, (2) the processor instruction set, (3) the address (and bus) width, and (4) the word size.

1. The *clock speed* determines the time horizon of CPU instructions. Each clock cycle accommodates one "atomic" CPU function. In some CPUs, a machine *instruction* only requires one clock cycle, or one "atomic" function; in other CPUs, a machine instruction may require multiple clock cycles, or several "atomic" functions. The microprocessor cycles are expressed in megahertz (MHz), or millions of cycles per second. Higher clock speeds result in faster execution of instructions. PCs installed by many organizations in the mid- to late 1980s operated at 8 to 12 MHz; these companies are now installing PCs operating at 20 to 50 MHz. Systems of increasing clock speeds are appearing as time goes on. For example, in 1992 Digital Equipment Corporation introduced a next-generation microprocessor (alpha chip) running at 200 MHz and capable of processing 400 million instructions per second (MIPS). This chip has 1,680,000 transistors.[12] CPUs operating at 250 MHz, supporting four instructions per machine cycle, and achieving 1000 MIPS are already available.[12] The instruction rate has increased almost logarithmically over the years; for example, the highest rate in 1981 was only 4 MIPS (see Table 2.3).

2. The processor's *instruction set* describes the set of "atomic" processes that can be carried out at each computing step, such as ADD, STORE, etc. A microprocessor can support 100 or more different instructions. Simple instructions require few processor cycles, whereas complex instructions require more cycles. Reduced instruction set computer (RISC) architectures, a relatively new development, limit (reduce) the number of composite instructions, thereby improving overall performance. RISC-based microprocessors are being deployed, particularly in high-end workstations.

3. The *address space* determines the amount of memory that can be directly accessed by the CPU. The bus is the physical facility that connects the CPU with the internal memory. If the number of wires in the bus is equal to the word size (see below), then data can be quickly moved from one module to the other. If the bus has fewer

TABLE 2.3   History of Key Intel Microprocessors

Name	Date of introduction	Clock speed (MHz)	Number of transistors	Applications
4004	1971	0.1	2,300	Calculators
8080	1974	2	6,000	Early PCs
8088	1979	8	29,000	IBM PCs and clones
80286	1982	12	134,000	PCs
80386 DX	1985	16	275,000	PCs and workstations
80486 DX	1989	25	1,200,000	PCs and servers
80486 DX2	1992	50	1,200,000	PCs and workstations
P5	1993	66	3,000,000	PCs and workstations

wires (typically one-half or one-quarter of the word size), then the word must be moved from module to module over multiple machine cycles, reducing the overall performance of the microprocessor. Hence, the width of the data bus—also known as the input-output (I/O) bus—determines the amount of data that can be transferred in one pass by the CPU. Intel's 80386DX or Motorola's 68030, which incorporate a 32-bit data bus, process information faster than older micros with 16-bit data buses. A 24-bit word microprocessor (and supporting bus)—for example, the 80286 and the 68000—allows for 16 Mbytes of memory (2^{24}=16,777,216). A 32-bit word microprocessor (and supporting bus)— for example, the 80386 and the 68020—allows access to 4 Gbytes of RAM. Many LAN servers use 32-bit microprocessors because of their ability to handle more data and increased input and output speed. A number of network operating systems (e.g., NetWare 3.11) require such microprocessors.

4. The *word size* refers to the size (in bits) of the data and instruction operands. The underlying hardware, registers, pins, etc. (but not the bus) must be consistent with the word size. Systems supporting larger words are more powerful: they can directly access more memory, and numbers can be represented with more precision. Many of today's LAN-based systems (workstations and servers) utilize 32-bit microprocessors.

Memory. This section discusses how *memory subsystems* are organized. Memory in a PC can be classified into two basic groups: (1) read-only memory (ROM) and (2) random access memory (RAM).

ROM is nonalterable memory that is used by the manufacturer to store a program such as Basic Input-Output System (BIOS), or other information. BIOS is a set of instructions that provide a means of communications between the microprocessor and such parts as the keyboard and the disk drive. BIOS can also be described as a set of routines that—in response to requests from the operating system or individual application—provides primitive control of the computer's devices such as disk drives, printers, video adapters, etc.[13] ROM memory is stored in ROM chips. In addition to traditional nonvolatile ROM chips, Erasable Programmable ROMs (EPROMs), allowing some degree of modification, are now relatively common.

RAM provides storage of real-time data needed to execute a program (some of this data is later stored in the "mass storage device," i.e., the hard disk drive). The CPU can write to and read from RAM. RAM is volatile, meaning that it cannot retain information without a power source and without a refresh cycle. Many programs in use today require 1 Mbyte or more of memory, typically up to 5 Mbytes. There are two types of RAM: Dynamic RAM (DRAM) and Static RAM (SRAM).

The DRAM capacitors store bits as an electrical charge. The presence of a charge represents a "one" bit (sometimes referred to as an "on" bit). The absence of a charge represents a "zero" bit ("off" bit). DRAM must be constantly refreshed to assure that the bits are not electronically discharged. This refresh process costs approximately 5 to 10 percent of the CPU's time. By 1995, 64-Mbit DRAMs should be available in volume production; 256-Mbit chips are expected in the late 1990s, and 1-Gbit by the early 2000s.[12]

SRAM does not require refresh cycles, but its complex circuitry makes it more expensive. SRAM stores bits in registers and therefore provides faster access time. When a bit is stored in SRAM, it retains a one or zero state until a change is made.

Cache memory is also being used with PCs to accelerate data retrieval. It is used for data that is accessed frequently: this data is stored in SRAM, thereby reducing accessing time. Data blocks of up to 64 kbytes can be retrieved in 35 nanoseconds (ns) or less.

Hard disk storage. Data that is required to run an application or is generated by an application is stored in RAM only temporarily. Eventually it needs to be permanently stored somewhere for future access. This is done by transferring the data to mass storage. A hard disk is the magnetic mass storage component of a PC, workstation, or file server which is permanently (or semipermanently) installed in the device. First-generation disks had a capacity of 10 to 20 Mbytes of storage. Consistent with the storage demands of today's applications, hard drives can now have a capacity of hundreds of megabytes. Magnetic disks now on the market are coated with iron oxide or a blend of iron. Approximately 50 Mbytes/in^2 of data can reside on these disks. Some vendors are starting to coat disks with composites based on aluminum. This aluminum composite results in densities of 160 Mbytes/in^2, implying capacities of 2.5 Gbytes on a 5.25-in disk and 1 Gbyte on a 3.5-in disk. Advances in manufacturing and data-encoding techniques, along with a competitive market, have resulted in decreased costs for mass storage in general and PC hard disks in particular. Also, writable—or, more precisely, write once read many (WORM)—optical disks are used by some organizations. Optical disks, which are mainly used for archival purposes, allow users to store from 600 Mbytes to 1 Gbyte of data on one disk.

2.1.3 Workstations

A workstation is a high-end device typically supporting a high MIPS rate, high-resolution monitors, and sophisticated applications. Most, if not all, of today's workstation vendors support attachment to an Ethernet LAN. Some workstation vendors support proprietary networks in order to increase data transfer speeds and enhance system addressing, which are not provided by Ethernet. Workstation connectivity is driving the movement to ATM-based LANs. Some put workstations in two classes: (1) technical workstations, and (2) diskless workstations.

Technical workstations.   A technical workstation is a microcomputer, usually possessing greater processing power than a PC. Technical workstations typically operate at 100 MIPS or higher; by comparison, traditional PCs delivered between 1 and 5 MIPS. Many vendors are now measuring performance in terms of millions of floating-point operations per second (MFLOPS). For example, the Sun Microsystems' SPARC 330 provides 2.6 MFLOPS. Many technical workstations use a floating-point coprocessor to improve speed and accuracy in numerical computation. Many workstations also support floating-point accelerators, which also improve processing speeds. Technical workstations usually incorporate a 32-bit CPU, large memory and storage capacity, and high-resolution monitors; they support the Unix operating system and the X Window System. Some high-end workstations include two or four multiple processors and operate up to 400 MIPS. High-end workstations range in cost from $15,000 to $55,000.

These workstations are designed for such applications as computer-aided design/computer-aided manufacturing/computer-aided engineering (CAD/CAM/CAE), electrical and structural engineering, artificial intelligence, imaging, technical publishing, and multimedia. Graphic processors, high-resolution monitors, and graphics applications support such services as visualization, imaging, and simulation with two- or three-dimensional views. The graphics applications often conform to industry standards such as the Graphics Kernel System, GSPC Core Proposed Standard Graphics Software System, and Programmer's Hierarchical Interactive Graphics System.

The graphics applications typically are accompanied by a windowing system and a database management system. X Window System interfaces provide multitasking for the applications residing on the workstation. Multitasking enables the workstation user to display multiple asynchronous events. Database management systems are used for data manipulation, applications development, and design retention.

Interest in RISC technology is increasing in order to achieve faster processing of information compared to the traditional microcode-heavy complex instruction set computer (CISC) architecture, found in many superminicomputer systems. RISC achieves higher speeds by implementing single-cycle execution. RISC is also more cost-effective than conventional architectures.

Diskless workstations.   Diskless workstations, as the name implies, are microcomputers which lack diskette and hard disk drives. Diskless workstations require a LAN: the operating system and application

software are sent to the workstation over the LAN. Diskless workstations offer a number of advantages over traditional PCs, particularly
in terms of software management. Also, these workstations are cheaper than the technical workstations or even PCs because they eliminate
the cost of drives. The initial installation of diskless workstations may
be met in some organizations with opposition from users. Users with
local confidential data may object to having their files stored in the
centralized repository. Such concerns may be alleviated by installing
file encryption software in order to assure needed confidentiality.

2.1.4 Supporting hardware

This section surveys some of the LAN-based hardware connected to
a LAN to support the computing models described in the previous
sections.

Cables. Considerations in selecting the type of cable include the (1)
medium (coaxial, twisted-pair, fiber, etc.); (2) physical characteristics
(typically with respect to local and national fire codes, plenum /non-
plenum, etc.); and (3) its electrical and optical characteristics.

   In a twisted-pair (10Base-T) LAN, which is now common, stations,
servers, bridges, routers, and gateways are attached to concentrators
using *two pairs* of twisted-pair cable. All devices are connected in a

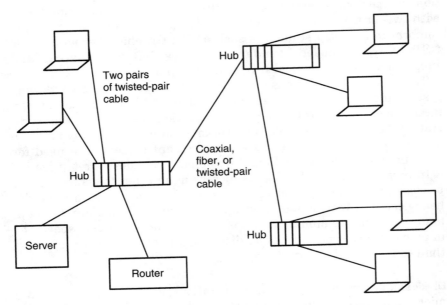

Figure 2.1 Twisted-pair (10Base-T) LAN wiring.

star configuration to a hub (discussed further below), which is usually located in a wiring closet. A hub (also known as wiring concentrator) supports from eight to a few hundred stations. Several hubs may be used to support a large population of users. In turn, the hubs are connected with each other, using any LAN medium (coaxial cable, twisted-pair, or fiber optic cable). Since there is a limit of 100 meters of twisted-pair cable from the hub, designers often use other media between the hubs to extend the distance (see Fig. 2.1).

Two physical characteristics of a metallic cable are the thickness and the outer diameter. In addition to physical characteristics, the cable plant can be characterized by the following electrical parameters:[14]

- Impedance
- Capacitance
- Attenuation
- Signal velocity
- Noise characteristics

Impedance. Impedance is a function of the resistance, inductance, and capacitance of the cable. The value should be relatively constant over a range of operating frequencies, to allow consistent and optimized signal transfer to the transceivers. For coaxial cable, this parameter is constant from 60 Hz to 10 GHz. For example, RG 11 cable has an impedance of 75 Ω; RG 62 cable has an impedance of 93 Ω; both the 10Base-5 and the 10Base-2 (described in Chap. 3) cables have an impedance of 50 Ω. For the twisted-pair cable this parameter is not constant and therefore the performance of the system decreases as the distance increases.

Capacitance. Capacitance is a function of the material (the dielectric constant of the material, to be specific), construction of the cable, and length. It has a direct effect on the signal quality by affecting the attenuation: the attenuation is directly proportional to the capacitance, which, in turn, is directly proportional to the cable length. In addition to attenuation there is also distortion (of the shape of the signal). After a certain attenuation-distortion threshold the receiver will not be able to adequately recover the signal. For coaxial cable the capacitance ranges from 13.5 (RG 62) to 26 (10Base-2) picofarads (pF) per foot; for twisted-pair the capacitance typically ranges from 11 to 17 pF/ft.

Attenuation. Attenuation is caused by the resistive and capacitative factors. It is directly proportional to the length of the cable. Hence, the lower the capacitance of the cable and the shorter it is, the lower will be the signal attenuation. Alternatively, one can lower the capacitance of the cable in order to extend its reach (at a given attenuation threshold). In LANs, a number of other factors control the ultimate length of the cable, including propagation delay and the medium protocol used.

Signal velocity.   Signal velocity is a function of the propagation medium. For example, in free space, electrical signals travel at 3×10^8 m/s. In other media, this speed is reduced. In coaxial cable, the signal velocity is from 0.78 (RG 58) to 0.83 (RG 11) times the speed in free space. For twisted-pair, the speed is 0.60 times the speed in free space. For fiber cables, the velocity is greater than 0.66 times the speed in free space. Signal velocity determines how quickly the signal can propagate along the cable; hence it determines the amount of time the protocol must assign to some event timers in its decision tables; in turn this determines the effective throughput.

Noise.   Noise characteristics are dependent on the type of cable (coaxial, twisted-pair, fiber, etc.) and the construction (shielded, unshielded). Noise is generated by external sources such as electric motors, generators, electrical power, fluorescent lights, and other nearby cable (in this last case the noise is known as crosstalk), and so on. Twisted-pair cable is the most susceptible medium, since it acts as an antenna. One way to reduce this is to increase the number of twists per foot.

Structured cabling.   Structured cabling has two advantages: it is application-independent, and it has a flexible, modular design. People and equipment can easily the relocated within an office (an event which is common in the corporate environment) without having to do rewiring. Growth is easy to manage. Maintenance is also easy, including fault sectionalization. In contrast, unstructured cabling may be application-dependent, often requires expensive rewiring, and does not support smooth growth or transition.

The Electronics Industry Association/Telecommunications Industry Association (EIA/TIA) has issued a standard for structured cabling in commercial buildings known as EIA/TIA 568. The standard covers four areas:

1. Medium (twisted-pair, coaxial, and fiber optic cable)
2. Topology
3. Terminators and connectors
4. Administration

The standard specifies the backbone as well as horizontal runs. Topology, distances, and administrative component are also covered. Such wiring systems are contemplated to have a useful life exceeding

10 years. The EIA/TIA 568 standard uses a hierarchical physical star topology. Logical LAN topologies such as bus, ring, tree, and dual ring can be mapped to this physical star wiring (see Chaps. 3 and 8). This hierarchical star enjoys the following advantages:[15]

- Supports a wide range of passive and active equipment
- Provides centralized points for managing and maintaining the network
- Provides a platform for modular and nondisruptive network growth

The standard partitions the wiring topology into five subsystems:

- Campus backbone subsystem with its main cross-connect (MCC)
- Building backbone subsystem with its intermediate cross-connect (ICC)
- Telecommunications closet subsystem (TCC)
- Work-area wiring subsystem
- Administration subsystem

The campus backbone subsystem links clusters of buildings that are in close proximity. One of the buildings in the campus contains the MCC in some convenient equipment room. The MCC consists of both active and passive components that support the backbone for the buildings in the campus which need to interconnected. Fiber cables are typically installed between the buildings when the company has its own right-of-way.

The building backbone subsystem (also known as a riser) provides the link between the campus backbone and the horizontal and work-group areas. This subsystem consists of an ICC located in some convenient equipment room. The ICC contains both active and passive components. Fiber cable is installed from floor to floor with a wiring closet on each floor.

The TCC, also known as the horizontal subsystem, provides the connection between the building backbone and the work-area wiring subsystem. The TCC is the boundary point between the riser fiber and other media. It consists of active and passive components. If the LAN system on the floor and work area in question does not require fiber, then the fiber terminates at the TCC. The fiber can be extended to the work area as the need arises.

The work-area wiring subsystem connects the PCs, MPCs, and workstations to the telecommunications closet. This subsystem typically consists of a communications box that can be configured to accept various cable connections.

The administration subsystem consists of the hardware needed to manage the subsystems listed above. It provides cross-connects (a patch panel consisting of a panel-mounted coupler and a patch cable to connect two cables from the same or different subsystems), interconnects (panel-mounted coupler to join two cables with a single pair of connectors), and documentation needed to achieve cohesive end-to-end connectivity across the network.

Network interface cards. PCs and workstations obtain physical access to a LAN with the use of a network interface card (NIC). A NIC is a "board" that is placed in the PC to support medium-sharing and timing functions. These cards typically implement the IEEE standards like 802.2, 802.3, 802.4, 802.5. Newer PCs and workstations come with integrated network interfaces. (See Figs. 2.2, 2.3, and 2.4.)

Transceivers. In addition to NICs, PCs may require a transceiver to link them to the LAN. The transceiver is a device that transmits and receives the signal between the PC's NIC and the actual medium (see Fig. 1.4). (Transceivers are not required in situations where Ethernet T-connectors are attached directly to a NIC's BNC connector). A transceiver can also be used to allow a connection to a medium other than the one native to the NIC. For example, a network interface card requiring a thick coaxial cable can communicate with a twisted-pair Ethernet network through the use of a transceiver. The transceiver provides a manager with added flexibility in configuring LANs with multiple medium types.

Figure 2.2 EtherLink III parallel tasking 16-bit twisted-pair network adapter (3C509 and 3C509-TP). (*Courtesy of 3Com Corporation.*)

Figure 2.3 EtherLink III parallel tasking 16-bit network adapter (3C509 and 3C509-TP). (*Courtesy of 3Com Corporation.*)

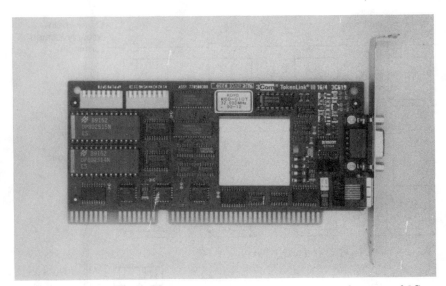

Figure 2.4 TokenLink III 16/4—16-bit ISA token ring adapter. (*Courtesy of 3Com Corporation.*)

Wiring hubs. Wiring hubs (discussed briefly earlier) provide a clustered control point for end-user device wiring. They provide the network administrator with a central cabling location to facilitate monitoring, moves, and growth. They also serve to link a LAN area (a group of offices, a floor, or perhaps even an entire building) into a larger backbone network. There is no industry consensus on what uniquely constitutes a hub (which initially was a simple concentrator): there is an "immense spectrum of available devices and features."[16] Many hubs support connections over shielded or unshielded twisted-pair cables, fiber optic cables, and thin and thick coaxial cables. Hubs typically support from 8 to 152 node connections, providing a network manager with a modular approach to network wiring and management. Hubs also allow a manager to control and monitor the network remotely, with the aid of network management software. Hub makers include Cabletron Systems, SynOptics, 3Com, and Ungermann-Bass. Hubs are being targeted by some vendors for upgrade to ATM by 1993. Key vendors see ATM technology and concepts as an important emerging force.[17]

There have been four generations of hubs.[18] Figure 2.5 depicts wiring evolution for Ethernet LANs. The traditional function of a hub is to extend and manage a physical network, as depicted in Fig. 2.6. Figure 2.7 shows an actual business example. In 1991 there were 2 million hubs installed in the United States; the number is expected to grow to 4 million by 1994 and 6 million by 1999.[18]

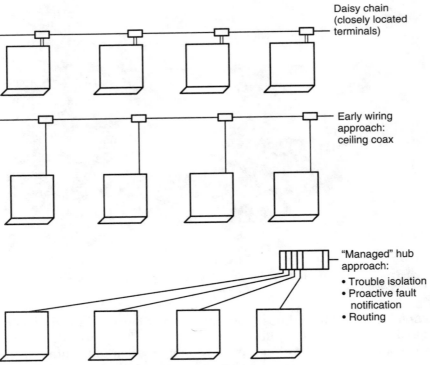

Daisy chain
(closely located
terminals)

Early wiring
approach:
ceiling coax

"Managed" hub
approach:
• Trouble isolation
• Proactive fault
  notification
• Routing

Figure 2.5 Evolution of wiring methods for traditional LANs.

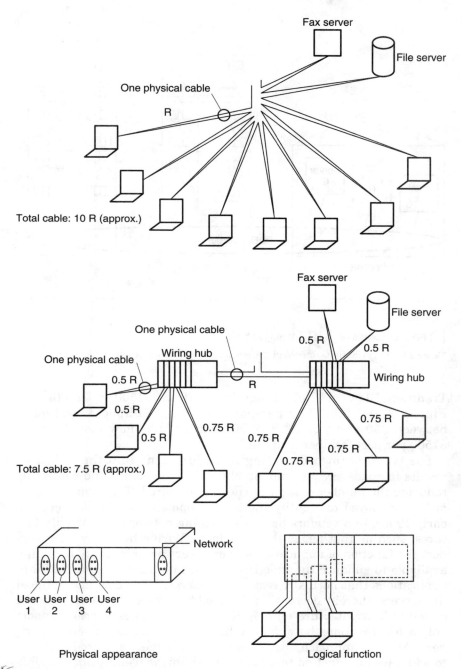

Figure 2.6 Advantage of wiring hubs.

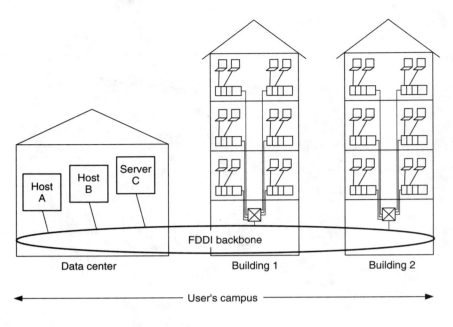

Figure 2.7 A typical first- or second-generation campus installation.

Traditional low-end hubs range in cost from $1500 to $2500. (See Figs. 2.8 and 2.9.) The per-port cost for a 10Base-T system is between $50 and $200.[19-21] (FDDI wiring hubs now cost $1000 to $1500 per port for fiber medium.)

The trend is toward putting more functions into the hub; this results in space savings and simplifies manageability and control by reducing the number of separate devices required. The term *managed hub* is employed to identify these new sophisticated hubs. Since the early 1990s, hub vendors have been adding network connectivity features, such as routing, to their products. Wiring hubs now also support the function of application servers. Newer products are becoming available to enhance the function of hubs into platforms for NetWare applications that normally reside on a server.[22] The goal is to off-load file servers, thereby decreasing the need to buy more. One issue is the closed-hub architecture employed by many vendors. These support only a few functions between the hub vendor and given server partners. Another approach, also seeing some commercial introduction, is to add a wiring function to a LAN server; this is usually only feasible for small LANs.

Figure 2.8 LinkBuilder 10BT and LinkBuilder 10BTi. (*Courtesy of 3Com Corporation.*)

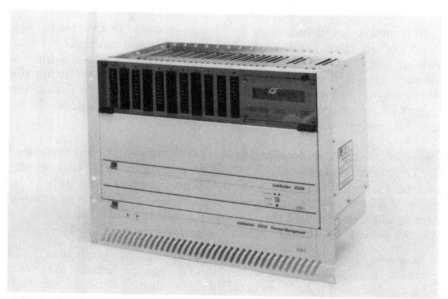

Figure 2.9 LinkBuilder ECS 10-slot Ethernet chassis hub. (*Courtesy of 3Com Corporation.*)

✳ Some of the new hub features include:

- Processing of alarms to determine if a problem can be solved without having to send notification to network administrator.
- Automatic start-up and shutdown at specified times.
- Ability to set network policies.
- Operation as terminal servers.
- Provision of LocalTalk to Ethernet gateways.
- Application platform: RISC processors with gigabytes of mass storage (file servers, print servers, directory servers) integrated in physically secure hub locations. Typical functions include routing, electronic mail, SNA gateways, etc.
- Provision of dedicated bandwidth to the desk (by pre-ATM means).
- Gateways (conversion and compatibility to other platforms).
- Integrated bridge-to-router support.
- Support of FDDI-to-token ring bridging.
- Security functions (port locking, packet scrambling, etc.).
- Provision of per-port switching.
- Hubbing for wireless LANs.

At a macro level, hubs can be categorized as (1) Ethernet hubs, fixed configuration, (2) Ethernet hubs, modular and multislot configuration, and (3) multifunction and mixed-LAN hubs. At the time of writing, about 30 vendors manufacture over 60 products fitting the first category; 15 vendors manufacture over 15 products fitting the second category; and 35 vendors manufacture over 37 products fitting the third category.[16]

✳ **Repeaters, bridges, routers, and gateways.**   There is an increased need for LAN interconnection, both locally and remotely. LAN connectivity can be extended using repeaters bridges, routers, and gateways. These devices support the connection function in a different manner and at distinct layers of the network's protocol hierarchy. Tables 2.4 and 2.5 provide a summary of functions of these devices.[23,24]

Repeaters.   Repeaters are devices that amplify the signals in order to increase the physical range of the LAN. A repeater usually extends a single segment of a LAN, to accommodate additional users, although some repeaters are capable of extending two LANs simultaneously. The two segments can conceivably use two different media or physical topologies, although this in not a frequent occurrence. Three limita-

TABLE 2.4   Comparison of Repeaters and Bridges

Repeater	Bridge
Operates at the physical layer	Operates at MAC sublayer
Available since late 1970s	Commercialized in mid-1980s
Cable systems connected to repeater are electrically dependent; noise may be amplified	Cable systems connected to bridge are electrically independent
Independent of higher-layer protocols (repeater type can be selected without concern for these protocols)	Independent of higher-layer protocols (bridge type can be selected without concern for these protocols)
Connects networks with same MAC and same upper layers	Connect networks with same LLC and same upper layers
"Protocol transparent"	"Protocol transparent": physical and MAC layers can be different on the two sides of relay; higher layers are the same on the two sides of relay
Stations are not explicitly aware of repeater presence	Stations are not explicitly aware of bridge presence, except in source routing
Does not affect address space	Unique address space (typically). Flat addressing; scan time proportional to number of active users
Amplification of all electrical and optical signals	Simple forward and filter decisions based on non-hierarchical address fields
No store-and-forward. Operates at the bit level. Default routing of all information.	Variety of routing methods (static routing, source routing, adaptive routing). Explicit protocols to disseminate real-time routing information between bridges not typically used
Simple installation, configuration, and operation	Simple installation, configuration, and operation
Trivial network management capabilities	Fair number of network management capabilities in the most sophisticated models
Local connectivity	Local and remote connectivity; best suited to implement an extended local network system
Simple cable-to-cable interconnection	Simple (spanning tree) interconnection topologies
No upper-layer protocol functionality (example error correction, readdressing, etc.)	No flow control; if packets arrive too quickly they are lost
Simple signal (bit) processing	Simple packet processing; fast compared to speed-preserving router but may affect medium speed (all other variables being equal, bridges will sustain several times the forwarding rate of routers; typical quoted rates: 6000 to 12,000 packets per second; actual rates: 40 to 60% of quoted figures)
Noise and distance limitation problems	Integrity problems (frame loss, misordering, duplication)
Cost: $1000 to $4000	Cost: $1000 to $20,000

tions of repeaters are (1) the amplification of noise, along with the signal; (2) the limited nature of the extension; and (3) the fact that the network remains a single network at the logical level, thereby keeping the number of users that can be supported bounded by medium-sharing considerations. They range in cost from a few hundred

TABLE 2.5   Comparison of Routers and Gateways

Router	Gateway
Operates at LLC and network layer	Operates at or above transport layer; typically at the application layer
Available since late 1970s	Available since early to mid-1980s
Subnetworks connected to router are independent	Subsystems connected to gateway are independent
In practical terms, the network layer tends to identify the communication architecture. Hence, router selection tends to depend on entire higher-layer protocol configuration	Highly dependent on communication upper layers and vendor
"Protocol specific": physical, MAC, LLC, and network layers can be different on the two sides of relay; in theory, transport through application layer is the same on the two sides of relay	"Protocol specific"
Routers are explicitly addressed by stations	Gateways are explicitly addressed by stations
Separate address space in different administrative domains	Separate address space in different administrative domains
Hierarchical addressing; scan time proportional to number of subnetworks in system	Hierarchical addressing
Complex internetworking arrangements including mixed technologies. Multiple types of network-layer services and protocols must be reconciled	Complex internetworking arrangements including mixed technologies. Multiple types of upper-layer services and protocols must be reconciled
Static (configured by system manager) or dynamic routing. Utilize explicit protocols to disseminate routing information between routers	Static (configured by system manager) or dynamic routing
More demanding installation, configuration, and operation	Complex operation, optioning, configuration, and management
Usually sophisticated network management capabilities	Fair degree of network management capabilities
Local and remote connectivity. Best suited to interconnect widely different LANs and LANs and wide-area networks	Local and remote connectivity. Best suited to interconnect widely different LANs
Complex interconnection topologies	Usually used in LAN-pair arrangements, rather than complex topologies
Flow control	Flow control
Relatively complex packet processing; performance becomes issue	Complex conversion at the communication level (layers 1 through 7)
Cost: $3500 to $20,000	Cost: $5000 to $20,000 (depending on complexity)

dollars to a few thousand dollars (for mixed-media repeating). (See Fig. 2.10.)

Bridges. Bridges connect two or more LANs at the medium access control (MAC) layer of a LAN. A bridge receiving packets (frames) of information will pass the packets to the interconnected LAN based on some forwarding algorithm selected by the manufacturer (for exam-

Figure 2.10   ISOLAN repeaters. (*Courtesy of 3Com Corporation.*)

ple, explicit route, dynamic address filtering, static address filtering, etc.). The receiving LAN must typically run the same MAC protocol as the transmitting LAN in order to read the packet (although translating bridges are also available—some administrators use routers instead for this function). As networks become more complex because of the addition of multiple departments and additional servers that share a common backbone, bridges provide the network administrator with the ability to divide the network into smaller logical segments to make them more manageable. Unlike repeaters, bridges regenerate the signals, so noise is not propagated. Over 50 vendors manufactured more than 100 bridge models at press time. Bridge products cost from $600 (for low-end basic local functionality), to $4000 (for midrange functionality), to $20,000 (for high-throughput, telecommunications-configured bridges). A typical system can be put in place for a few thousands of dollars.

Figure 2.11 depicts typical possibilities for local bridges. Segment-to-segment bridging is relatively straightforward. Multiple cascaded segments are possible; however, the traffic destined for a "far" node must pass through several bridges, causing a possible degradation in quality of service (delay and frame loss). Multiple LANs use backbone bridging to avoid local traffic congestion, although a first- or second-generation backbone may eventually become a bottleneck. Multiport

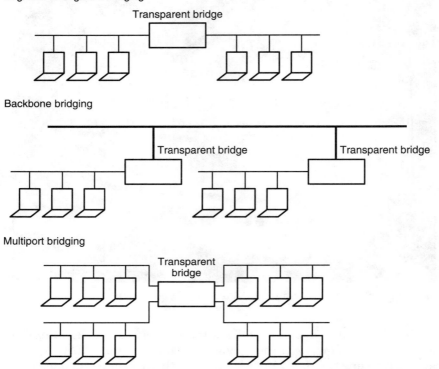

Figure 2.11 Typical bridging strategies.

bridging allows several LANs to share the bridge. Devices communicate with each other through the bridge's internal bus. Shared resources and improved network management can reduce the per-port cost of bridging.

Vendors are adding features to bridges on an ongoing basis. Some enhancements being made at the time of writing include larger address tables, complex frame filtering, increased throughput, load balancing, wider selection of WAN interfaces, redundancy and support of network management capabilities (including standardized protocols). Many vendors (close to one-third) are also adding variable levels of routing (these devices are known as *brouters*), allowing users to deploy bridges to interconnect two or more LANs and later invoke routing capabilities as additional protocols are added to the network.[25] Routing support is for traditional upper-layer internetworking protocols, as well as for other international standards.

"Transparent" bridges utilized in Ethernet (i.e., bridges which do not require the user to specify the path to the destination) need to maintain address tables. Until the early 1990s, these tables were able to store 2000 to 5000 entries; newer bridges can store up to 60,000 entries (15,000 to 20,000 entries are more typical). In contrast, "source-routing" bridges used in token ring LANs require that the sending user station (i.e., the source) supply instructions as to how to reach the destination.

A transparent bridge's address table is maintained on a static or dynamic basis. In the static environment, the LAN administrator specifies whether or not a frame for a given destination needs to be forwarded to the downstream LAN. In a dynamic environment, the bridge builds its own table by "observation." Every bridge in the network must maintain a table with entries for all active users. Because of table entry limitations, the bridges employ aging techniques to delete destination entries with no recent activity. If a frame arrives at the bridge for a destination which is not on the table, a flooding technique is used. This, however, impacts the overall performance of the bridge. Hence, a larger address table is advisable.

Frame filtering and forwarding rates vary from a few thousand per second at the low end, to a few tens of thousands at the midrange, to a few hundreds of thousands at the high end (as high as 200,000).[25] Rates in the 10,000 to 20,000 range are common.

Bridges supporting FDDI are also entering the market, as well as bridges using wireless channels for the bridge-to-bridge interconnection. The feature contrast between bridges, routers, and hubs is now blurring in product implementations. Integration of bridge, router, and hub functions is achieved by loading appropriate software to underlying hardware platforms, or by adding a bridge-in-a-card or a router-in-a-card to a wiring hub.

Routers. Routers connect at a higher protocol layer than bridges (at the network layer). Routers provide flow control for the incoming LAN packets, thereby increasing the reliability of the interconnection, and allow the use of a variety of interconnection subnetworks. Different packets can, in principle, be routed over different networks, for example, for security or least-cost-routing reasons.

Routers operate with a particular WAN protocol or with a number of protocols. If multiple protocols are being used to interconnect LANs, a manager can either select a separate router for each protocol or have a router that is capable of retaining multiple protocols in one chassis. Disadvantages of routers, relative to bridges, include reduction of packet filtering speed and increased cost. However, low-cost routers—$3500 range—are beginning to emerge.

Gateways. Gateways are used to interconnect LANs that employ completely different protocols at all communication layers. The complete translation of incoming data units associated with completely different protocols affects transmission speed.

Servers. A server is a LAN-attached microcomputer-based device that provides functions such as file storage, printing, or communications services to other entities on the LAN. File servers which are used to retain a repository of shared subdirectories for LAN users require mass storage capabilities. As networks grow and support higher levels of data-intensive applications, the server performance becomes critical to overall performance.

A *superserver* is a server that incorporates multiple processors and is equipped with multiple ports for peripheral attachment. The processors of a superserver, usually two to eight, allocate various tasks in order to expedite processing of user requests. Superservers, which started to appear in the late 1980s, can be used as database servers (most typical) and application servers. Typically, the superserver refers to a workstation that is not designed to be a user PC, one that has fast processing speed, supports multiprocessing, and has extra storage (say 90 Gbytes). They range in price from $20,000 to $100,000.[29]

Communication servers. The increased need to access off-net services, including companywide e-mail, fax services, public-access databases, file transfer, and remote printing, have led to expanded use of communication servers. These servers provide a variety of communication functions. The most basic function is interface to the public network. This is done with a modem for analog services, and data service units for digital services.

A *modem* is a device that allows a digital signal (such as the one generated by a PC, a bridge, a router, etc.) to be transmitted over an analog communication facility. Modems support speeds in the range of 9.6 to 38.4 kb/s (with compression) and are priced competitively. Features, such as autodial, auto-answer, and self-diagnosis are common on many modems. Some modems, such as those based on the CCITT V.42 *bis* standard, support data compression. Modems were once viewed as isolated communication units, but they now operate with LAN management systems in modem pool arrangements. A *modem pool* is a communication device with multiport and multipoint capability that eliminates the need for separate modems at the user's location; alternatively, it can be a (small) collection of discrete modems centrally located—these may be rack-mounted or not. Modem pools provide LAN users with multiple modems for dialing remotely, on a first-come, first-served basis. The LAN users establish connections to the modem pool through communications software.

Modem pools are typically used on LANs where the amount of network traffic does not warrant a bridge or router on a dedicated communication line and/or the destinations are widely dispersed. A plethora of digital services is now becoming available, including ISDN, T1, switched T1, frame relay, SMDS, and cell relay; these services do not require modems (see Chap. 7). While small companies and low-end LANs will probably continue to use modems, large companies and high-end LANs will likely migrate to high-speed digital services in the next couple of years.

2.1.5 PC operating systems

This section refocuses on PCs on a LAN, but from a software perspective. An operating system (OS) is a software program that manages the hardware and software resources of a computer. Typically, it is associated with a given PC hardware platform. Changes to operating systems were slow in developing. Changes between versions have typically been spaced years apart. The introduction of faster microprocessors has led to more rapid changes in operating systems. Microsoft Disk Operating System (MS-DOS) currently has the largest installed base. However, the market for multitasking—the ability to simultaneously do two or more things, as provided by OS/2 and UNIX—is starting to expand. The changes in operating systems have resulted in higher speed and support for large files, along with multitasking and interoperability features.

MS-DOS. MS-DOS is a single-user, single-tasking operating system which originally supported only 640 kbytes of memory. This amount of memory is marginally adequate for many if not most of today's applications. Microsoft's newer versions of DOS are capable of supporting larger amounts of memory. The addition of Microsoft Windows gives MS-DOS users the ability to achieve multitasking. Windows, often referred to as an operating environment, is an application with a graphical interface that allows user access to more than one application. It also allows the transfer of information among the accessed applications. Windows and applications supported by Windows are being implemented at a rapid pace on DOS-based network operating systems.

OS/2. OS/2 is a single-user, multitasking operating system developed jointly by IBM and Microsoft. OS/2 supports preemptive multitasking. This implies that the operating system is sharing the CPU with other active applications. These active applications do not have the memory constraints of MS-DOS. The 80286 microprocessor can access up to 16 Mbytes of real RAM, compared to MS-DOS's (basic) 640 kbytes. Two important features of OS/2 are interprocess communications and dynamic data exchange. Interprocess communications allows the OS/2 user (or applications) to communicate with each other and share data. The dynamic data exchange provides links between applications. This feature lends itself to distributed processes with the ability to broadcast services across a network. The OS/2 operating system can operate with an Intel 80386 microprocessor. (OS/2 2.x series is a 32-bit operating system.)

Unix.   Unix is a multiuser, multitasking operating system. AT&T Bell Laboratories developed Unix over 20 years ago. AT&T (System V), BSD (Berkeley Software Development), and Microsoft (Xenix/386) are typical Unix operating systems vendors. Unix, like OS/2, takes advantage of Intel's 80386 microprocessing capabilities. Unix is used with many high-end workstations.

Macintosh System 7.   Macintosh System 7 is a multitasking operating system developed by Apple Computer. The Macintosh operating system is based on the Motorola series of microprocessors. Features of this operating system, such as the graphical user interface, which uses windows, icons, and a mouse, were first explored by Xerox PARC. The graphical user interfaces (GUIs) aim at ease of use. The graphic capabilities of this operating system have made it a choice for desktop publishing applications. Multitasking features have been simplified with the graphical, iconic-based displays of GUIs. Apple Computers pioneered the commercial introduction of GUIs. Software vendors have recognized the merits of GUIs and are providing graphical front ends for their products.

2.1.6 Network operating systems

NOSs are software facilities that support and control multiuser access to disk drives, printers, and other peripherals on a LAN. They intercept and redirect requests for service to the appropriate server. Hence, an NOS is networking software that enables the network to support multiuser, multitasking capabilities. Although a LAN could function at a basic level without a NOS, the NOS facilitates resource sharing and effective communication. The NOS determines the class of devices that may be attached to the LAN and arbitrates users' service requests. Some of the more commercially prominent NOSs at this time are Novell's NetWare, Microsoft's OS/2 LAN Manager, Banyan's VINES, 3COM's 3+Open, AT&T's LAN Manager/X, and IBM's LAN Server.

NetWare has the largest installed base of network operating systems today. The newer versions of NetWare operate on servers using the Intel 80386 or 80486 microprocessor. Add-on software enables NetWare servers to be accessed by PCs using Macintosh, Unix, OS/2, or DOS operating systems. Banyan's VINES is a NOS that is structured using the Unix operating system. VINES is adaptable to networks with many users and large volumes of traffic. VINES also offers network managers interoperability and ease of adding, deleting, and moving users. 3COM's 3+Open supports disk caching, disk mirroring, multiprotocol support (including TCP/IP, XNS, and OSI), centralized administration and security, multiprocessing, and peer services such as file and printer access. Microsoft's OS/2 LAN Manager, AT&T's LAN Manager/X, and IBM's LAN Server are NOSs that use a server engine using OS/2. LAN Manager is available from many "original equipment manufacturers" (OEMs).

2.1.7 Logical LAN architectures

Beyond the basic traditional NOS environment, two "new" logical architectures have emerged for developing cooperative distributed applications now being deployed in LAN-based environments: client-server architecture and peer-to-peer model. These architectures utilize both the OS and the NOS to support the tasks associated with distributed computing.

Client-server architecture. A client-server LAN architecture is a computing model in which software applications are distributed among entities on the LAN. The clients request information from one or more LAN "servers" that store software applications, data, and network operating systems. The network operating system allows the clients to share both the data and applications that are stored in the server and the peripherals on the LAN. Figure 2.12 depicts a logical view of a client-server environment, while Fig. 2.13 depicts a physical view. Note that multiple clients can rely on a single server. Approximately 200,000 client-server configurations were forecast to be in service by 1993, although actual deployment may in fact be less.[26]

There are three basic functions supporting computing: (1) data management, (2) processing, and (3) presentation (to user). Client-server models allow the distribution of these functions among appropriate devices as shown in Fig. 2.14. Table 2.6 depicts some of the features associated with client-server "implementation" variations.[27,28]

Any system in the network can be a client or a server. The client is the entity requesting that work be done; a server is the entity performing a set of tasks on behalf of the client. The user's processor con-

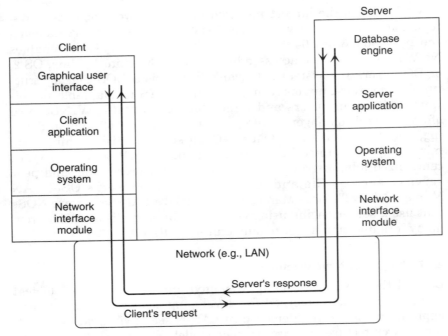

Figure 2.12 Client-server model—logical view.

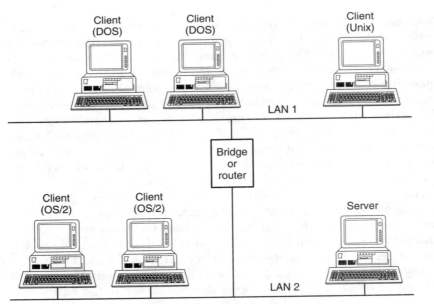

Figure 2.13 A physical view of a client-server configuration.

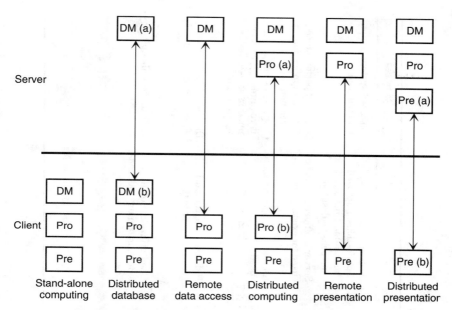

Figure 2.14 Various client-server implementations. DM = data management. Pro = processing. Pre = presentation. xxx(a/b) = function xxx partially done.

trols the user interface and issues commands to direct the activity of the server across the LAN. This is done through the use of remote procedure calls (RPCs). RPCs are software programs with distributed capabilities. Applications that are implemented on the LAN can "call" these procedures by ordering messages, translating different codes, and maintaining the integrity of the protocol. Not all applications in a client-server architecture are stored on a server: clients are capable of storing applications and data locally. When clients possess individual operating systems, the network is referred to as *loosely* coupled. Some of the benefits of a client-server architecture include:[27]

- Increased productivity
- Control or reduction of costs by sharing resources
- Ease of management through focusing efforts onto a few servers
- Ability to adapt to needs

In the stand-alone computing model, all the intelligence is placed in the PC. Even when the data resides in a file server, the file server simply stores the file in a functionally equivalent manner as if it were a local hard disk. For example, to do a sort, the entire file is downloaded to the PC, which does all the computing; the file may eventually be reloaded in sorted form to the server. The server has no ability

TABLE 2.6 Features Associated with Client-Server "Implementation" Variations

Variation	Application	Examples	Pluses	Minuses
Distributed presentation	Existing mainframe-based applications	PCs; X-terminals; X-servers	No changes to existing applications; improved human-machine interface; inexpensive desktop	Increased system computational load
Remote presentation	Independent management of user environment; new or existing applications	Workstations; PCs; DEC window/Motif	Server is offloaded; can be used for existing applications	Degraded performance for multiapplication environments
Distributed computing	Computing components executed by most appropriate platform (array processors mainframe, mini, etc.)	Workstations; high-end PCs; minicomputers; mainframes; adjunct processors	All resources are optimally used	Cooperative computing is required (processing must in some sense be coordinated)
Remote data access	Common data with independent data applications; decision support systems	Workstations; high-end PCs; minicomputers; mainframes; SQL access; fourth-generation languages; DECquery	Reliable data; computing choices closer to user	Database machine performance affects all users
Distributed database	Resource sharing for desktop applications	Workstations; high-end PCs; minicomputers; mainframes; Oracle's Oracle Server; IBM's Database Manager; Sybase's SQL Server; Novell's NetWare SQL; Gupta's SQLBase; other fourth-generation languages	Good utilization of all devices; maximum independence	Difficult to scale up

to manage or control the data. There are two problems with this static server architecture.[29]

- The number of simultaneous users is rather small (≤ 10) owing to performance problems related to the movement of files and indexes.
- Multiple applications usually require replication of the data into multiple databases, often of different file formats. This makes database synchronization difficult.

In a client-server environment, the file server has the ability to perform database management. This means that the server (also known as the *back end* or *database engine*) can run a relational database management system using a multitasking operating system (e.g., OS/2 or Unix). See Table 2.6 under "distributed database." A relatively powerful microprocessor (e.g., 80386-level) is required to run the multitasking OS, the database manager, and the concurrent user sessions.

In a client-server environment, the workstation (also known as the *front end*) is responsible for "presentation" functions (i.e., displaying data according to specified user interfaces, editing and validating data, and managing the keyboard and mouse). These functions are easy to implement, making the cost of the repetitive module (n modules for n end users) low. This results in the following computing advantages:[30]

- Support of many concurrent users (a few hundred)
- Reduction in cost, since front-end software is simple
- Software from a variety of vendors, performing a variety of functions and sharing the database
- Data integrity (single database shared by all users)

The four common methods of linking client-server applications are application program interfaces (APIs), database servers, remote windowing, and RPC software.

Application program interfaces. A commonly used method of sharing information on a client-server network is through the use of network APIs. These are vendor-provided functions that enable application programmers to access network resources in a "standardized" manner. Network APIs also allow for the connection of various applications running under the same operating system. Vendor-supplied APIs are not easily portable among different operating systems. SPX (Novell NetWare), Named Pipes (Microsoft LAN Manager) and Sockets (Berkeley Unix) are examples of APIs. A long-term goal is to come up with vendor-independent APIs.

Database servers. A database server is a dedicated server in a client-server network that provides clients with distributed access to database resources. The database servers usually employ a structured-query language (SQL) relational database to communicate between the client and the server. Using SQL, database servers allow a client to download a table, as opposed to downloading the entire database. This feature reduces LAN traffic, thereby improving performance. SQL database servers extend user processing applications across various network operating systems via RPCs. SQL is a simple conversation-like language relying on English commands such as SELECT, FROM, WHERE, etc. to perform database inquiries. For example, the user can issue the command

SELECT customer, balance FROM checking-account-list
WHERE balance ≤ 1000

Remote windowing. Remote windowing is an extension of the "windowing" concepts commonly used with PCs. Remote windowing allows for multivendor connectivity through the use of the universal terminal standard. This concept allows the viewing of multiple-user processing applications concurrently from remote locations. In order to properly distinguish Microsoft Windows from the generic concept of windowing, the following definition is provided. *Windowing* is a software feature that offers split-screen capability in which the different partitions of the display form rectangular areas. These rectangular areas are usually accompanied with GUIs that can be moved or resized on the screen. Microsoft Windows is a specific vendor application providing windowing.

Remote procedural calls. RPCs are based on computer-aided software engineering (CASE) principles. This concept allows conventional procedure calls to be extended across a single-vendor or multivendor network. RPCs function across several communication layers. Programmers are shielded from the networking environment, allowing them to concentrate on the functional aspects of the applications under development.

The computing industry generally describes two RPC technologies as industry standards. They are Open Network Computing (ONC) RPC (often called SunRPC), and Network Computing System (NCS). An international standard RPC based on the Remote Operations Service Element and Abstract Syntax Notation One is under development, with possible completion in 1993 (ISO Project JTC1.21.48).[31]

A 1991 survey of 1600 customers found that 22 percent had already implemented client-server architectures; 27 percent may implement them in the near future; and 51 percent were not planning to implement them.[28] Typical applications include office automation, use of communication servers, and transaction processing.

Software vendors experienced some difficulties in migrating from a file server to a client-server model. Releases took longer than expected to reach the market.[30] It seems to have taken longer than expected for user programmers to become acclimated to the new environment. Nonetheless, migration is under way, and by 1996 client-server architectures will be common.[30]

Peer-to-peer model. Peer-to-peer networking represents a transition away from the traditional mainframe-based network strategy, where the mainframe treats user devices as "dumb" terminals endowed with limited or no capabilities. The peer-to-peer model injects distributed principles to support direct communications among users without having to necessarily rely on the mainframe for routing, as is the case in a traditional hierarchical network.

Early LAN products included IBM's PC LAN Program and DCA's 10Net. Peer-to-peer has had its genesis in SNA's move to a distributed architecture, although the case could be made that open-interconnection data communication standards have been the real drivers. In an IBM SNA mainframe environment, advanced peer-to-peer networking (APPN) implements the peer-to-peer model between any appropriately configured set of devices. These devices are capable of undertaking the routing and control functions that were previously centralized in the host (see Fig. 2.15). (In fact, APPN is now part of IBM's networking blueprint for migrating SNA to a distributed architecture.)[32]

Early peer-to-peer LAN products required a large amount of PC memory, were often tied to proprietary LAN hardware, and offered limited features.[33] More recently, peer NOSs have evolved to utilize 80386-based servers and to use standard LAN hardware (typically Ethernet). With memory managers and improved NOSs, networking can now be accomplished in a straightforward way. Observers see products now being ideally suited to penetrate low-end networking in small companies.[33] Typical products include Invisible Software's Net/30, Hayes Microcomputer's LANstep, Artisoft's LANtastic, Mainlan's MainLAN, WEBcorp's WEB, D-Link System's LANsmart, Novell's NetWare Lite, and Performance Technology's POWERLan. Products like LANtastic, POWERLan, and Net/30 are full-featured and belong to the high end; there is a midrange and a low-end category.

The best peer-to-peer NOSs now support good performance, provide powerful features and flexibility, and afford ease of management. These products are generally targeted to smaller installations; there are environments where high-end server-oriented NOSs (e.g., LAN

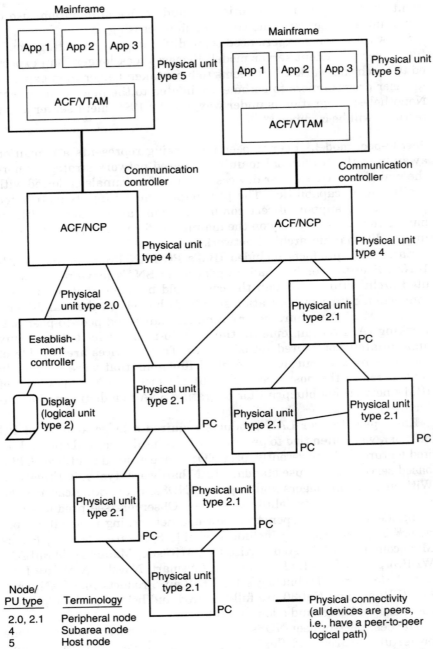

Figure 2.15 Peer connectivity among multiple processors communicating within the same network and between networks.

Manager, VINES, NetWare 3.x, etc.) are the only reasonable choices. Advantages of peer-to-peer NOSs include:

- Flexibility. Every PC can become a server, or a single PC can be designated as a server (as is the case in a traditional LAN, for example under NetWare); PCs can also easily switch roles between being a user of services and being a provider of services.
- Ease and simplicity in installation, management, and use.

Peer-to-peer NOSs still had a few limitations at the time of writing; notably, most use proprietary protocols, limiting interconnectivity. For example, no router support is available to interpret the LAN's protocol layers above the MAC (bridges can, however, be employed). Additionally, LAN analyzers which can decode the protocol fields are not available. Figure 2.16 provides a guide for selecting a traditional NOS and a peer-to-peer NOS.

2.2 LAN Interconnection

As the popularity of PCs and LANs grew in the mid-1980s, problems related to interworking the various departmental platforms become more pressing. Users on different "islands" still had a need to share data, reports, and word processing documents, just as they had done in the days of centralized computers. Interconnection of LANs has increased over the last few years owing in part to new technologies and the requirement to distribute business functions. These distributed business conditions originate with the need for organizations to carry out their business operations at multiple locations, both domestically and internationally. As a result, organizations are seeking to link their local LANs with remote LANs in order to extend consolidated databases and other applications to workers at remote sites.

The price of LAN solutions typically places them within the decision-making level of the department head. In turn this implies that various departments may make independent decisions, without initial concern for interdepartmental connectivity. Connectivity needs, however, soon become manifest to all involved. In an increasing number of organizations, the responsibility for connecting these "network islands" into one enterprisewide system is shifting back to a central organization like the data processing department.[34]

When considering the issue of connectivity beyond physical bridging and in terms of higher-layer protocols, two common solutions that have emerged are based on Novell's IPX/SPX protocols (for local interconnection) and TCP/IP protocols (for wide-area interconnection).

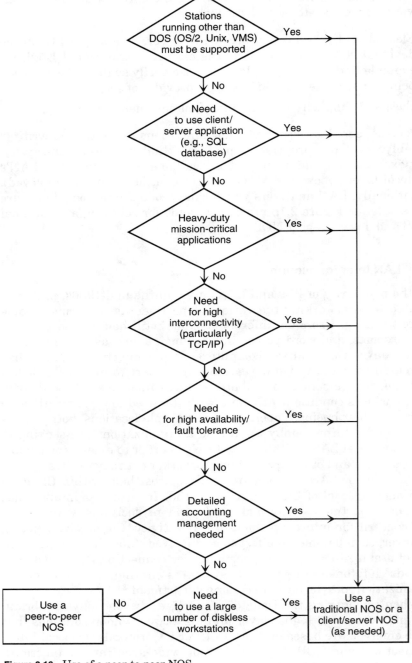

Figure 2.16 Use of a peer-to-peer NOS.

These are now considered de facto standards. Open-system interconnection (OSI) protocols are also emerging, but progress is not as rapid as some had expected (OSI is discussed in more detail in Sec. 2.4). Figure 2.17 depicts these and other protocols commonly in place.

At the hardware level, physical connectivity is accomplished using bridges, routers, bridge-routers (brouters), and application-level gateways discussed earlier. Fairly similar platforms are connected with relatively inexpensive and fast bridges. Platforms which have some differences, particularly in terms of how they deal with subnetworks, need more expensive and slower routers. Routers are used when multiple networks can be employed, and addressing, flow control, and LAN-to-WAN protocol matching, adaptation, and conversion is required. The "multiple networks" can be as simple as two physically different routes (dedicated lines), or a dedicated line with overflow to the public network service at the busy hour (such as switched T1), or the ability to choose a WAN service such as X.25 packet switching, frame relay, or SMDS. The choice can be determined by the network manager through programmable parameters. The route (or network) can be the least expensive route, or the highest-capacity route, or the route with the least traffic. Routers are used locally in lieu of bridges to guarantee flow control between the LANs or if protocol conversion above the logical link control layer is required (this is discussed in Chap. 3). Completely different environments require yet more expensive and slower gateways. Figures 2.18 and 2.19 depict some examples. Figure 2.20 depicts a typical router backbone network supporting connectivity between remote locations, using public frame relay services.

2.2.1   Local interconnection

Local interconnections involve the linking of LANs within the geographical limits of a building. Campus interconnections link two or more buildings that are in close proximity to each other. Wide-area networks connect LANs that are dispersed over greater distances.

As LANs grow in terms of the number of users, more information is passed through the shared physical network media. At some point, the level of traffic on the LAN will cause overall end-to-end performance to degrade. Some networks experience degradation during periods of heavy application use. A local bridge has the ability to logically and physically partition the network. This has the effect of load-balancing the traffic into separate segments. Without a bridge, traffic originating anywhere on the network can appear on all network interface cards (i.e., at all stations) connected to the LAN. A filtering bridge eliminates nonlocal traffic and can enhance performance on

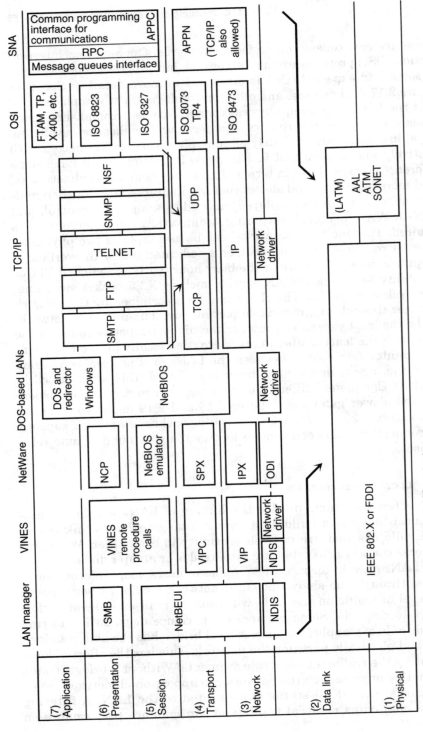

Figure 2.17a Plethora of protocol suits in LAN context.

SMB   Server message block (a distributed file system enabling application on one PC to access a file elsewhere on the network as if the file resided locally; analogous to NCP and NFS; developed by Microsoft, IBM, and Intel).

NetBUEI   Net BIOS extended user interface (Microsoft's version of NetBIOS).

RPC   Remote procedure calls (lets system call a remote service).

VIPC   VINES interprocess communications protocol (transport layer protocol that supports both "best-effort" transmission and "reliable" transmission).

VIP   VINES Internet Protocol (VINES's internetworking protocol, similar to IP).

NDIS   Network driver interface specification (device driver specification developed by Microsoft and 3Com).

NCP   NetWare core protocols (distributed naming service under NetWare).

NetBIOS   Network basic input/output system (the software layer that links the NOS to the hardware).

SPX   Sequenced packet exchange (a set of NetWare commands that operate above IPX to ensure data delivery).

IPX   Internet packet exchange (NetWare's network layer protocol).

ODI   Open data-link interface (NetWare's device driver).

DOS   Disk operating system.

Redirector   A software module that intercepts requests from the applications to the operating system and routes them across the network.

SMTP   Simple mail transfer protocol.

FTP   File transfer protocol.

TELENET   Terminal emulation and communication program.

SNMP   Simple network management protocol (see Chap. 5).

NFS   Network file system (a distributed file system; developed by Sun Microsystems and used by a variety of UNIX workstation vendors).

TCP   Transmission control protocol (see Chap. 4).

UPD   User datagram protocol (see Chap. 4).

IP   Internet protocol (see Chap. 4).

FTAM   File transfer, access, and management (ISO's transfer protocol).

APPC   Advanced program-to-program communications (SNA's peer-to-peer upper layer protocol).

APPN   Advanced peer-to-peer networking (IBM's distributed transport/network layer protocol).

LATM   Local ATM (see Chap. 9).

Figure 2.17b   Acronyms and concepts for Fig. 2.17a.

each segment. Routers are principally aimed at WAN connectivity but can also be employed locally.

2.2.2   Internetworking of multiprotocol platforms

As discussed earlier, it is not unusual for companies to have multiprotocol LANs operating alongside SNA networks, that is, to have equip-

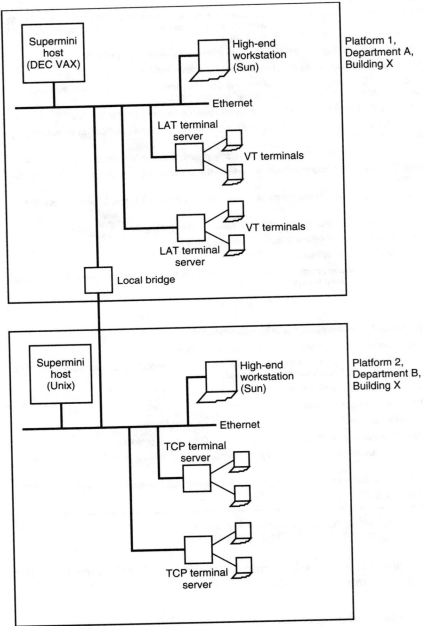

Figure 2.18 Use of local bridges in environments of similar platforms.

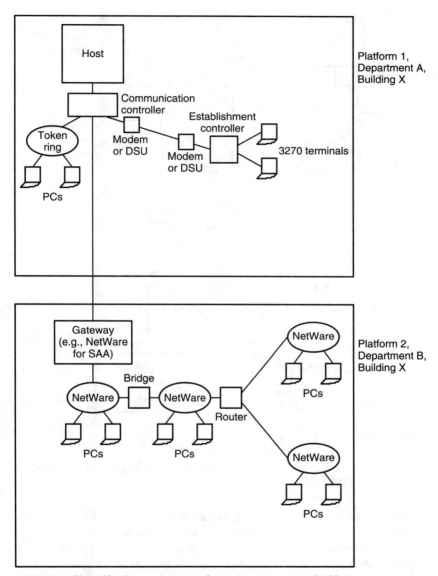

Figure 2.19 Use of bridges, routers, and gateways to connect LANs.

ment connected as shown in Fig. 2.14. There are about 8 million 3270 terminals in U.S. corporations, and millions of staff hours have been invested in applications written to operate in the 3270 data stream environment.[34] This scenario, however, can be expensive, because of the duplicate backbone, particularly when the network is regional or

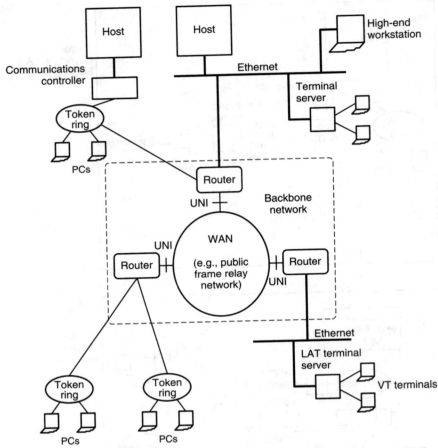

Figure 2.20 Backbone network for connecting multiprotocol environments. UNI = user-network interface.

national in scope. Consolidating these networks not only reduces transmission costs but also reduces the cost associated with administration and maintenance.[32] The challenge is then to move SNA traffic over the LAN-based backbones. A variety of bridge and router companies has announced a number of strategies for doing this, including SDLC tunneling and source-route bridging.

SDLC tunneling transmits 3270 PDUs by adding Internet Protocol headers to the 3270 data stream for transmission over the backbone network, as seen in Fig. 2.21 (Internet Protocol is described at length in Chap. 4). The SDLC PDU is encapsulated by the bridge and routers. The encapsulation is added at the local end and removed at the remote end.

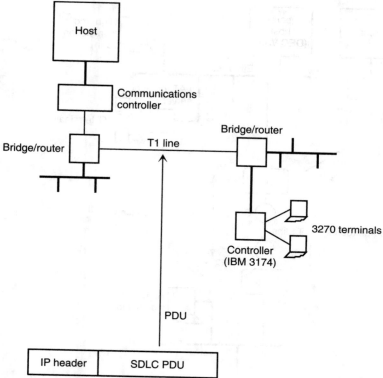

Figure 2.21 IP-based encapsulation.

Source routing is a method for bridging token ring networks, including LANs, in the SNA environment. Source routing requires that the source of the message specify the routes which the message must take over the backbone network (in transparent routing, the equipment automatically forwards the message to the destination). A number of bridge and router companies are proposing to add source routing methods to routers supporting Ethernet-based protocols, to run concurrently with these. By adding support for source routing to their equipment, bridge and router companies targeting the SNA market are able to support Ethernet and token ring traffic simultaneously (see Fig. 2.22). Work is under way to complete an IEEE *bridging standard* that allows direct interconnection of Ethernet and token ring networks. These use an algorithm called "source routing transparent," which is a hybrid between transparent bridging typically used in Ethernet bridges and source routing typically used in token ring networks.

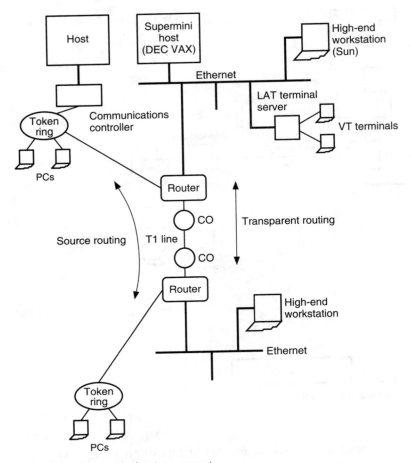

Figure 2.22   Source routing transparent.

2.2.3 LAN interconnection over a distance

Higher wide-area network speeds are needed to support high-speed second- and third-generation LANs and their data-intensive applications. This WAN traffic can consist of a few kilobits to several megabytes of data. Bandwidth requirements will continue to grow, especially as multimedia applications and imaging applications become more prevalent.

One way to reduce overall networking costs (i.e., to do global, corporatewide optimization) is to consolidate multiple wide-area networks. Teletraffic considerations show that by combining two hitherto separate networks into one, the manager can provide the same grade of service to the users while reducing the number of communication links. This process entails deploying a backbone network shared by the various users and applications. In fact, throughout the 1980s companies have moved in this direction.[23]

There are several WAN approaches for interconnecting LANs. Dedicated bridge-to-bridge or router-to-router links can become expensive when the length of these links is large (nationwide or international links) and when there are many LANs to be interconnected. Public packet-switching services based on X.25 protocols have the advantage of being widely available from the carriers. Support equipment is also available. Packet-switching service offers economic advantages for dispersed LANs with moderate bandwidth and modest latency requirements. The drawback of X.25 packet-switching service is low throughput and relatively high end-to-end delay; these limitations are particularly restrictive when LANs support new, data-intensive applications. New WAN services offer a high-speed interconnection based on cell relay technology. Cell relay services (discussed in Chap. 9) provide high bandwidth connections with low latency for applications such as imaging, multimedia, and other distributed, data-intensive applications. (This use of the term "cell" should not be confused with the use in the context of the distributed computing environment—DCE—discussed later.)

The Internet research community is currently studying the problems of supporting a distributed applications over interconnected networks.[35] These applications, which include multimedia conferencing, data fusion, visualization, and virtual reality, require that the distributed system (the hosts and servers that support the applications along with the interconnected network to which they are attached) be able to provide guarantees about the quality of communication between applications. For example, a video conference may require a certain minimum bandwidth to be sure that the video images are delivered in a timely way to all recipients. One way for the distributed system to provide guarantees is for hosts to negotiate with the interconnected network for rights to use a certain part of the internetwork's resources. An alternative is to have the interconnected network infer the hosts' needs from information embedded in the data traffic each host injects into the network.[35]

The issue of LAN interconnection will be treated in more detail in later chapters (Chap. 7 in particular).

2.3 LAN-based Applications

2.3.1 Traditional business applications on LAN-based PCs

This section briefly surveys some of the typical business applications currently deployed on LANs. While next-generation LANs are already available and are being deployed, these LANs are likely to be initially used for high-end scientific and supercomputer applications. Traditional LANs will probably be the mainstay of business for at least another decade. Typical business applications and application trends are discussed below. LAN managers of Fortune 1500 companies will surely be required to continue to support these applications and design effective LANs which support production staffs using these applications for day-in and day-out business functions.

LAN-based applications of the mid-1980s started out with limited features. Business applications now provide sophisticated features such as messaging functions, basic graphical functions, support for on-line conferencing, and direct access to remote applications through the use of windowing functions. These applications provide integrated tools that combine traditional data processing with graphics, image processing, and, in some cases, sound and voice processing. The goal of the application vendors is to provide consistent interfaces to all applications and platforms. Some applications are beginning to support the integration of text, audio, and video functions; however, this is typically done locally (at the PC, for example, with a CD-ROM reader), without networked servers supplying this data or real-time worker-to-worker sessions because of the bandwidth limitations of first- and second-generation LANs.

E-mail applications. E-mail continues to be one of the most widely used LAN applications. Traditionally this has involved text, but multimedia messages composed of multiple "body parts" (e.g., voice, graphics, fax, video, etc.) are beginning to be supported by newer e-mail applications.

Database applications. Many business functions involve database access and update. Examples include customer service functions, reservations, ordering and processing, records update, and so on. Many LAN-based database operations now use SQL. An SQL database, working in conjunction with a database server, can reduce the amount of real-time traffic on a LAN, affording better overall response time between the server and the client. When a client needs information, a request is made to the SQL server, but rather than giving the client a copy of the entire database, the server sends only the subset that is needed.

Spreadsheet applications. An increasing number of corporate users need financial data to carry out their functions, such as sales inquiries, expense information, statistics compilation, and aged-receivables analysis. This type of analysis is now typically done with a PC-based spreadsheet application. Often such analysis requires data stored in other PCs or servers. Additionally, the resultant data may have to be sent electronically to the intended user or to another coworker for review and enhancement.

Project management applications. Mainframe project management applications have been used for many large projects such as civil engineering and government projects. Their main function is to estimate the time of individual tasks and projects. Today many project management, scheduling, and tracking applications are located on LAN-based PCs. The output consists of printed reports and charts; the data can also be stored and backed up in a server.

Desktop publishing applications. Advances in PC technology, along with low-cost laser printers, have made it possible for the transformation of traditional publishing functions. Desktop applications for corporate word processing typically require LAN networking in order to print files, share information, or perform joint editing and composition functions.

While most corporations have some form of desktop publishing, as just described, there are industries, such as the publishing industry itself, which are heavy users of desktop publishing applications. Publishing, in this context, is the function of disseminating information to the public using the print media (e.g., newsletters, magazines, books, newspapers, catalogs, advertisements). The print media are still the most widely used communication tools for conducting business and reporting news, and reading is still one of the most popular forms of entertainment. Print media are not expected to change to an electronic presentation format until the next decade. The publishing industry has secured gains in productivity using computers to automate the work flow in the publishing process for magazines, books, newspapers, and advertisements.

In the 1980s, PCs led the migration of parts of this process to the desktop in smaller satellite offices of newspapers or to individuals' homes. Publishing has merged with computers. This permits the work flow in the publishing process (e.g., writing, editing, graphic arts, typesetting, and printing) to be automated and made more efficient. Newer network communication services like frame relay and SMDS provide the capability to gain even greater efficiencies by supporting distributed operations.

Electronic publishing entails the use of computers and telecommunication systems to enable the production and distribution of a printed product through photographic composition in a cost-effective and rapid manner. The photographic composition includes the new technology associated with multimedia, image capture, and transfer.

Newsprint organizations face competition from information service providers (e.g., Prodigy) capable of delivering news to the home electronically. Some of the newspaper chains are already experimenting with electronic delivery of newspapers in different parts of the United States. The driving force for this change can be attributed to three major developments:[35]

1. The availability of 16-bit and 32-bit PCs with the power of midrange computers. These systems support graphic, image, and text software packages, at a relatively low cost.
2. The availability of laser printers for creating graphics that produce results similar to typography and the emergence of color printers, at a relatively low cost.
3. Standards for images, graphics, and print formats and the standardization of the interface between text and typesetting formats.

The combination of these developments permits the rapid creation of text, graphic, and image data by different people on a LAN. These files can be shared and edited by workers in an expedited manner and can be printed locally for further processing and inspection.

Graphics and presentation applications.   Graphics can be divided into three major categories: business, presentation, and data analysis. Business graphics transform data into charts and may include object-oriented drawing capabilities. Presentation graphics also include tools for data-driven charts and object-oriented graphics but allow for a greater range of free-form capabilities. Data analysis graphics provide a user with graphics from large quantities of numerical data. Often this data is transferred over a LAN, either to a server or, if nothing else, to a printer.

Many desktop publishing applications, including simple word processors, allow copying and pasting of graphics. Again, this data is circulated over a LAN for either local output (to a printer) or to a remote user on an interconnected LAN.

Integrated applications.   Apple System 7, Microsoft Windows, and OS/2 support integrated software. Integration, in this context, refers to the combination and sharing of information among applications. The integration process combines the different applications under a uniform interface. The user can easily switch from one application to another; data can also be moved among applications. These systems use windowing techniques supported by icons.

Microsoft Windows provides integration without using add-on integration applications. OS/2 allows a user to perform multitasking cut-

and-paste data transfer between applications. This is accomplished by permitting the cut-and-paste functions to take place in the background. OS/2 Presentation Manager goes beyond the functions of cut-and-paste by establishing links between applications. (Apple System 7 also supports a similar function.) Thus, the update of one application will be automatically carried to another application. The linking of various applications is accomplished with the use of a mechanism called dynamic data exchange (DDE). The utilization of DDE is widespread. Microsoft DDE products, such as Word and Excel, are popular.

Software development applications. Many of the business applications of the 1970s and 1980s that were developed for the mainframe and midrange computer systems required experienced in-house programmers. Users now rely on "canned" applications purchased from software vendors. In some cases, however, tailored applications are still being developed in-house. Some users need specialized applications that are not available from vendors. It may not be profitable for the vendors to create applications for a small community of interest. The alternative is to provide the users with developmental applications that will satisfy their immediate needs. This user "development" is normally done over a LAN.

As a special case, the software development industry develops applications which can have as many as 500,000 lines of code. The use of LANs to support application development is widespread for the obvious reasons of backing up code, cooperative work and development, printing, downloading code developed locally to a faster machine or debugger, etc.

In the PC environment, object-oriented programming (OOP), compilers, and interpreters can be utilized not only by experienced programmers but also by novice users and casual programmers. An object represents a collection of data and behavior. The object also represents information concerning its nature and functions. OOP, when properly implemented, can reduce software development and software maintenance time.

CAD/CAM/CAE applications. CAD/CAM/CAE is the process of using a computer system to create, visualize, dimension, simulate, and test the performance of a product prior to the physical process of manufacturing it. Often the selection of components or materials and the product description accompanies each design. Computer-aided design and engineering has evolved in many companies from a single terminal connected by a host to a network of workstations or PCs on LANs sharing libraries of designs and drawings, as well as databases and peripherals. Additionally, the design centers of major manufacturers (e.g., automobiles, aerospace, etc.) are often distributed across the country. The need to electronically connect design centers with dispersed manufacturing factories is on the rise.[36]

Computer-aided manufacturing is the process of using a computer to control the manufacturing process (e.g., numeric control of machining, process control in chemical production, assembly line sequencing, etc.). CAD/CAM/CAE systems often use a cooperative processing architecture (i.e., client-server). This architecture employs a central system for the storage of application designs. Designs are then downloaded from the server to a workstation or PC, as needed. The input devices include workstations or PCs, typically with 32-bit processing and high clock speeds, graphic tablets, mice, light pens, and scanners. The output devices include plotters, printers, 19-inch high-resolution displays, and computer output microfilm (COM). When the server and the client are remotely located, a suitable communication service is required to connect them.

Graphic applications which go beyond basic objects (pie charts, bars charts, graphs, etc.) require large amounts of memory. For example, 25 Mbytes of storage is required to digitize an $8\frac{1}{2} \times 11$-in color image using a 16.7 million color palette. The amount of memory required grows exponentially with the number of colors used. This is one of the reasons for wanting to move these applications to third-generation LANs.

2.3.2 New applications

Table 2.7 depicts some of the new and evolving scientific and business applications. These applications may have been tried and prototyped on traditional LANs, but the only way to make them practical is to deploy them over a third-generation LAN.

Imaging applications. Electronic imaging is rapidly emerging as an important business application of the 1990s in the medical, banking, and scientific communities, to list a few. Imaging integrates digital image data from text, graphics, and video sources. There is an increasing market for digital imaging services. Common services include the transmission and storage of high-resolution medical images, transfer of financial and insurance images, the accessing of electronic library services, and the sharing of graphic designs. Image transfer occurs when the image data is moved from one storage or display device to another. Typical image transfer applications include medical imaging, engineering product design, secure image analysis for defense, and insurance claims processing. Third-generation LANs will probably be the vehicle for efficient delivery of such data to the desktop.

TABLE 2.7   Evolving Data-Intensive Applications of the Early to Mid-1990s

Computer-aided design/computer-aided manufacturing/computer-aided engineering (CAD/CAM/CAE)

Computer-supported cooperative work (e.g., joint document editing, cooperative design)

Computer-aided communication

Telenavigation (travel agency applications: CD-based in early 1990s, networked in late 1990s)

Image transfer (health care—picture archiving and communications systems)

Data-intensive process control, including LAN-based robotics

Tele-education (also known as virtual college or distance learning)

Desk-to-desk videoconferencing

Multimedia conferencing

Multimedia messaging

Inter-enterprise electronic data interchange

Network computing ("distributed computing environment")

Virtual or artificial reality*

    Networked applications for remotely controlled robotics, exploration, etc.

    Entertainment and games

*Speculative; late 1990s at best.

Electronic imaging is the process of (1) obtaining or creating digital image data from a paper document, graphic, or video source; (2) performing optional encoding of the data; and (3) storing it in electronic or optical memory. The stored data can be accessed by a computer and directed to an output device for visualization (i.e., to a display device or a printer).[36] Some of the business advantages of electronic imaging systems compared to current procedures include reduced storage costs, reduced cost of distribution, and increased productivity.

Image transfers typically require a large bandwidth for movement of files representing x-rays or other scientific images (up to 4 Mbytes). Examples of imaging and image transfer devices that may be located on a LAN include flatbed scanners, optical character recognition (OCR) systems, facsimile machines, gray-scale and color printers used in office automation or engineering applications, medical x-ray scanners, x-ray film printers used in a hospital or medical application, digital video cameras, and red-green-blue (RGB) color monitors used in entertainment applications.

These applications are driving the introduction of third-generation LANs and "gigabit" WANs. Table 2.8 depicts some typical throughput for a single user-to-server transaction. (This may have to be multiplied by n to support n users.)

TABLE 2.8   Imaging Data Rates

Resolution	Colors (expressed in bits/pixel)	Size of an image (Mb)
256 × 256	6	0.4
512 × 400	24	4.9
1024 × 1024	12	13
1024 × 1024	24	25

Multimedia.   PCs and workstations are now being upgraded to support multimedia, specifically simultaneous presentation of text, audio, and video. Multimedia requires appropriate platform software; it also requires hardware such as CD-ROM drives, sound cards, speakers and microphone, and, possibly, musical instrument digital interface (MIDI) support. These PCs are known as multimedia PCs (MPCs). Sounds can be edited; bit-mapped graphics can be edited, merged, and placed in motion; graphics and sounds can be synchronized. The dynamic nature of multimedia interactions makes this technology ideal for training purposes.

Initially, multimedia applications are supported by access to a nearby (stand-alone or server) CD-ROM. Eventually, this multimedia information will also be located remotely. Examples include:

- General or technical library
- Tele-education material provided by a university or learning center
- Corporate information or training located at a central site
- User-to-user multimedia real-time conferencing or multimedia conferencing over a great distance

Existing services (e.g., frame relay) may be inadequate to handle LAN interconnection to support multimedia; other services (e.g., dedicated T1 and T3 circuits) may be too expensive. Multimedia client-server architectures or groupware applications can generate high-intensity traffic. Workstation vendors are now pursuing ATM as the technology of choice for these evolving applications. ATM facilitates effective multiplexing over a WAN of streams supporting different media. ATM is also upgradable in terms of the underlying speed (1.544 Mb/s, 45 Mb/s, 155 Mb/s, 622 Mb/s, 1.2 Gb/s). ATM-based hubs, supporting star architectures, allow tributary devices to be connected at their own speed, within each cluster, without having to require that all devices operate at the highest speed (as is the case in an FDDI environment).

Geographic information systems. A geographic information system (GIS) is an information system that is designed to work with data referenced by spatial or geographic coordinates. A GIS is both a database with specific capabilities for spatially referenced data and a set of operations for working with the data. Such systems tend to be data-intensive.

Tele-education

One day in the life of an adjunct: Until now.* It's another Wednesday at 4:30 p.m., and tonight, after eight hours of regular work, you head north for your adjunct teaching assignment. You immediately dish out 25 cents in tolls and drive 20 minutes with traffic, in darkness, while you eat a sandwich with one hand. Dish out 35 cents and drive another 30 minutes. Then dish out $1.85 in additional tolls. Drive 5 minutes in heavier traffic, then pay up another $4 in additional tolls. Eight lanes converge into a two-lane tunnel; 10 minutes later you're in the city. Fight the aggressive window washers, pushy cab drivers, jay walkers, street peddlers, and other characters, and 15 minutes later you're in the parking lot. Dish out $12 for prepaid parking, pack two attachés (one for normal papers, the other for lecture papers), maybe try to hold an umbrella in a windy and rainy night, go one block out of your way to get some much-needed (but, turns out, lousy) coffee for $1.25 and placed in a weak paper bag, walk back two blocks and at just about 6:10 you're in front of 20 eager students.

Lecture for three hours in sometimes overheated, sometimes underheated rooms, with or without chalk readily available. At 9:15, start the journey back home, retracing all the steps above, including the tolls, plus 20 more miles since you live south of where you work (and pay up additional tolls). It's 11 and you walk in the door, after a round-trip drive of 120 miles. Catch a cold bite and hit the bed by midnight. Do that for years, sometimes twice a week.

One day in the life of an adjunct: Now. What a dream, after 10 years! It's 4:30 p.m. and you head home to teach. At 6 you have supper, then play with the kids for a while. Go to the study to read some trade press, perhaps a news weekly, and around 8:30, with some Vivaldi in the background, you dial into the local public frame relay node, which lets you connect at 64 kb/s with the university's LearningLAN and its KnowledgeServer. You first check to see if there is e-mail for you from the administration or other members of the faculty. Next, you send some mail to the department's chairperson about some grade-related

---

*If the reader does not appreciate a "slice-of-life," he or she should skip over the first few paragraphs of this subsection.

matter. Then you log on to the main LearningWare program and start opening up files from each student to determine how the last assignment was handled by each of them. Some assignments involved teams of several students. Students sit comfortably in their offices, at their LAN-based desktop, or at home, and, at a time of their choosing, also log on to the university's KnowledgeServer while listening to their individually selected music and having set the ambient temperature to personal taste. Users can be in 50 states or even on several continents.

You individually annotate each assignment, giving detailed textual responses. Now Respighi is playing in the background. Questions that the student might have asked the instructor are individually answered. Correct or "orthodox" answers can be "pasted" into the file directly from "canned" files residing on the server; text, graphics, scanned photos, sounds and speech, and view-graphs are included. Even an instructor's spontaneous verbal comment can be sent (at 1000 bytes per word: 500 ms of speech per word, at 16,000-b/s encoding). A one-minute comment translates into 120 kbytes and can be sent in real time over the 64-kb/s line to the student's mailbox on the server, which can store a variety of e-mail body parts. After returning the individually annotated "papers" to each student, all the assignments are stripped of the author field, merged, and sent, with a single command, to all the students for reading.

It's 9:55 p.m., Paganini is playing, and it's time to start this week's lecture. With a single command, the textual, graphical, and voice-encoded information comprising the evening's lecture (uploaded once onto the KnowledgeServer and just slightly updated before the term starts) is moved from your folder in the server to the folders of all students, giving them instruction, direction, and assignments. Assignments must be completed by no later than "next Wednesday." It's now 10 p.m.; you log off the KnowledgeServer and you turn on CNN for the day's events. A couple of days later, maybe during lunch at work, you get a jump-start and log into the KnowledgeServer from your LAN-based desktop to see if any student has asked any questions that may preclude him or her from continuing or completing the assignment.

At the end of the semester it's a breeze to give grades. Additionally, all sorts of statistics as to students' connect time, preferred session time, etc., are readily available. All projects, now in electronic form, particularly the good ones, can be stored for future use in other classes.

The university has also "sold" an individualized version of the course to company XYZ. A group of employees is asked to spend two hours a week, Tuesday 9 to 11 a.m., to take the course from their

LAN-based desktop systems. You follow exactly the same procedure for the matriculated students, but do that on Saturday morning.

One day in the life of an adjunct: Tomorrow. The "now" approach described above pales by comparison. Real-time n-way multimedia communication (i.e., multimedia conferencing) is now a routine event. All students see you and you see all the students, wherever they are. In addition to real-time instruction, you have precanned clips (as in a TV studio) of graphical, pictorial, verbal, and audiovisual material which can be "pasted" into the outbound stream with trivially simple icon and object commands. Also you have access to a real-time library of facts, information, definitions, etc., which can be searched with any number of indexes and delivered on an interactive basis as needed.

The real thing: NYU's Virtual College.[*] The spread of computers and telecommunications and the rise of global markets have rendered traditional bureaucracies increasingly unwieldy. Hierarchical organizations are being replaced by more flexible work groups whose participants are located across a city or an ocean, and whose offices have become not literal physical spaces but "virtual" workplaces—that is, a series of networked computers. Tomorrow's managers and professionals will need both design and work within electronic environments that connect people as well as computers.

These changes are reflected in the growth of telecommuting—the use of communications and computers to perform work outside of the traditional workplace. Out of 5.5 million telecommuters in 1991, 14 percent were executives or managers and 10 percent were professional specialists. Such major organizations as American Express, J. C. Penney, AT&T, General Electric, Pacific Bell, and Travelers Insurance have instituted extensive telecommuting programs to hire and retain talented, but remote, staff and to reduce office operating costs.

NYU's Virtual College is an on-line training network that helps prepare managers and professionals for the new organizations of the 1990s and beyond. Using their home PCs and modems, the college's students receive instruction, ask questions, conduct analyses, resolve problems, and prepare professional studies—all largely at their own pace and convenience. The Virtual College is an electronic learning environment for the efficient production and delivery of a wide range of high-quality business and technical courses.

---

[*]This section was contributed in its entirety by R. P. Vigilante, director, New York Information Technologies Institute.

As the physical infrastructure of international business is changing from concrete and steel to computers and networks, the Virtual College gives students the collaborative and technical skills for working within (as well as on) today's decentralized and electronic workplaces—in effect, a virtual classroom preparing employees for tomorrow's virtual organizations. The program employs Lotus Notes software.

- *Lotus Notes.* Lotus Notes is a powerful group communications program giving people who work together an electronic environment within which to create, access, and share information, using networked PCs. Lotus Notes supports such business applications as computer conferencing, information distribution, status reporting, project management, and electronic mail.

- *Computer requirements.* Students must have an 80286 or 80386 IBM-compatible PC with Microsoft Windows 3.0, 2 Mbyte of memory, 8 Mbyte of available hard disk space, 1.44 Mbyte 3.5-in diskette drive, VGA monitor, and a 2400- or 9600-b/s Hayes or Hayes-compatible modem. Optionally, users can access the telecourse from a work location using a LAN-based PC which either has a direct dial-out capability or uses a communication server (modem pool).

Corporate training is a $100 billion-a-year business in the United States. Upwards of 35 million individuals receive formal, employer-sponsored education each year. As an example, over 1 million professional employees of CPA firms are required by law to complete 40 CPE credit hours of technical training annually.

The facilities, staff, and curriculum costs of traditional corporate training programs are considerable. If the wages of trainees are included, these costs double. To meet the continuing demand for training, education has to be constantly available to employees through convenient and economical means. The cost-effectiveness of on-line versus on-site training is of increasing interest to thousands of business and public organizations.

The largest audience for on-line training is the nation's 32 million managers and professionals. According to a new study by the Census Bureau, 17 million (56 percent) of these individuals regularly use computers at work and 10 million (33 percent) have computers at home. Over 3.1 million home computers (23 percent) have modems. In 1991, the potential on-line training market base of managers and professionals in the United States, with modem-equipped home computers, is approximately 2 million people.

As technology continues to transform organizations and markets, managers and professionals need training to take on more responsibility, sharpen analysis and judgment, and improve communication skills. NYU's Virtual College and its on-line educational network are designed to meet these expanding training needs in an efficient and effective fashion.

The Virtual College courseware packages are developed on an integrated PC-based production system. Incorporating the latest desktop publishing and computer graphics software with digital video and laser image-setter hardware, the system efficiently produces a wide set of instructional materials (see Fig. 2.23). Thoughtful design procedures will ensure maximum flexibility in instructional materials utilization across various media. The same still video frame might be an image in a hypertext database and an illustration in the accompanying textbook. With minimal modifications, the same computer graphic will be used in a database disk, videocassette, textbook, and color slide. Updating text and images will largely be done once, and not separately for each medium. This all-digital production system will generate timely and high-quality courseware for NYU's Virtual College.

NYU's Virtual College delivers training to students by a series of interactive multimedia packages. Each course's package will consist of a textbook, a hypertext database (on magnetic or optical disks), and the data network access software (see Fig. 2.24). The textbooks will provide the basic subject content for each course. Many terms and concepts in the text will be "linked" to identical but more detailed and

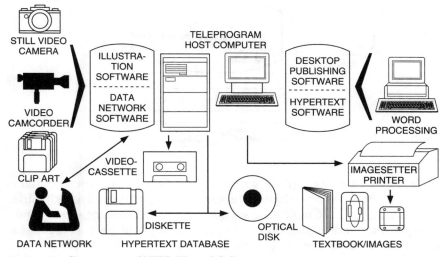

Figure 2.23 Components of NYU's Virtual College.

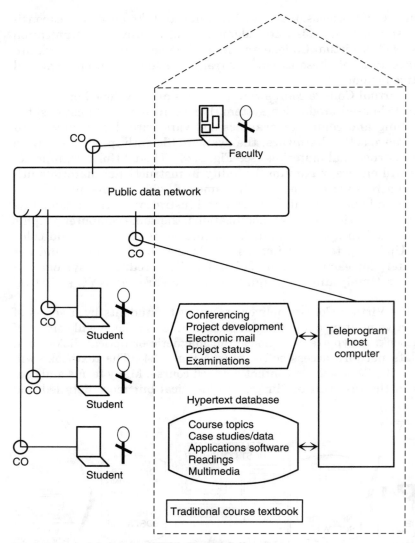

Figure 2.24 A global view of NYU's Virtual College.

interactive references in the hypertext database. The hypertext database will bring course topics, techniques, and technologies to life, allowing students to nonlinearly explore concepts, readings, case study simulations, and application software. Throughout each course, students will dial up the program's data network to discuss topics, ask questions, submit assignments, complete projects, and take examinations. The data network will permit student-faculty interactions and simulated problem solving to occur in real time, and not be limited to the artificial weekly instructional increments of traditional courses.

A hypertext database organizes information in a nonlinear format that supports active cross-references and permits the user to "jump" to various parts of the database as desired. Interacting with the database through associative links, users can follow their individual trains of thought and nonsequentially access text, graphics, data, software, and video materials at varying levels of detail.

Each telecourse has a hypertext database containing files of course notes, case studies, applications software, data sets, and required reading. At appropriate points of the course, students access selected database files and interact with them on their personal computers. In the example of Fig. 2.25, the student is reading the textbook chapter on input design. Coming across a highlighted cross-reference (called a link) to Jones, the student can, if interested, access the hypertext database for Jones's complete article on user interface design. Within

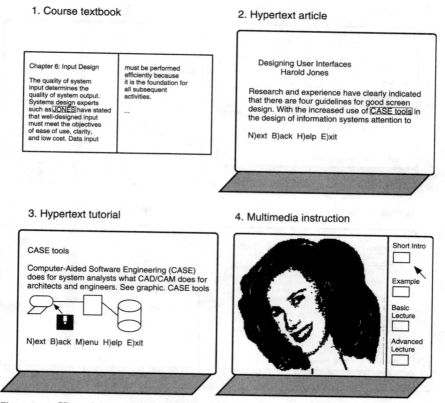

Figure 2.25 Virtual College instruction material.

the Jones article, a link to CASE tools permits the student to call up a graphic tutorial on computer-aided software engineering tools. Finally, students with PCs having digital video and optical disk features will be able to access video demonstrations, graphic animations, and computer simulations, enhancing their understanding and mastery of program topics.

The data network will provide students and faculty with a menu of computer conferencing, project development, electronic mail, and other on-line instructional services (see Fig. 2.26).

Computer conferencing will provide for direct interaction between students and faculty. Many topics in the program have ongoing computer conference that require meetings and analyses related to the implementation of case studies, projects, and assignments. Each computer conference generates a considerable and permanent record of faculty-student discussion and analyses.

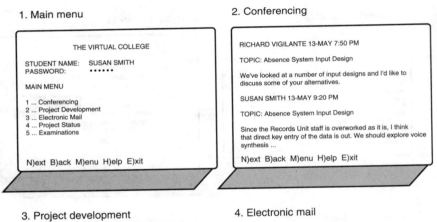

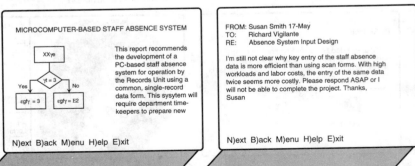

Figure 2.26 On-line facilities in support of the Virtual College.

Project development allows work groups of students and faculty to go beyond discussion to the creation of tangible group projects (e.g., a marketing plan, an information system, a manufacturing design). Students create professional-quality printed reports, electronic presentations, and computer models and simulations—all while quite distant from one another.

Electronic mail provides students and faculty with the ability to compose, send, and receive assignments and readings. A major use of e-mail will be to send program assignments to faculty and return corrected assignments to students.

Network computing/distributed computing environment. We consider distributed computing environment (DCE)—the new term for distributed data processing—an application. (In this discussion, an application is any software-based task requiring computing and communication services.)

Users are now seeking GANs supporting interoperability among heterogeneous computing platforms. The goal is efficient, flexible, secure, and inexpensive distributed data processing capabilities, where the processing power and information is allowed to be dispersed over a network of many interconnected computers. Network computing, another name for DCE applications, may entail transacting small, medium, or large files across LANs, WANs, and GANs. It typically employs a client-server model. Existing networking systems (including NOSs) fall short of some of the key functions required to make the goal a reality.

The DCE software is being developed by the Open Software Foundation (OSF), an industry-supported organization with about 350 worldwide hardware and software vendors, end users, and research institutions. With DCE, users will be able to develop, deploy, and manage applications that efficiently use networked computing resources. The DCE focuses on four interconnection problems.[37]

1. The diversity of operating environments.
2. The magnitude of the problem, in terms of the number of computers that need to be interconnected. (Existing systems such as NetWare 3.11 can accommodate networks of hundreds of systems; evolving networks of thousands or tens of thousands of systems require new software.)
3. The need for security.
4. Scalability: the need for growth and for new network applications.

File sharing within an organization became common through software such as NetWare for MS-DOS PCs and Sun Microsystems Inc.'s

Network File System (NFS) for Unix workstations. However, sharing of data has remained relatively localized within the LAN because of de facto incompatibilities, not only in the communication protocols (for example, a PC with a token ring card cannot communicate directly with a PC with an Ethernet card), but more fundamentally at the operating system level.

There are some who seem to believe that the problem is solved as soon as two physical "legs" of a connecting network are signaled to be connected, providing an end-to-end bearer service. This simplistic communication example works when one is trying to establish a pipe through the network in order to connect an "ear and a mouth" (E&M)—an obvious pun for those who recognize the E&M term—but is not sufficient or adequate in the data communications context. Such physical connectivity is just the beginning of what it takes to make two data devices communicate. Communication has been defined as follows:[38]

> A and B are communicating when A is suitably able to code a message of information, and can relay it to B through an appropriate medium, and B is suitably equipped to receive the message *by using or interpreting it in some fashion.*

Simply establishing a physical pipe between two devices is like providing the air in a room (i.e., the medium over which voice can be carried) where two individuals, one speaking only Ugaritic and the other only Phoenician (languages dating from 2000 B.C.), are brought together; clearly, no intelligent communication can take place. As datacom practitioners, sometimes we have difficulty conveying to others what is meant by the term *data communication,* or *datacom, expert.* A datacom specialist undertakes all the tasks, functions, hardware deployment, software deployment, protocol compatibility matching, and network establishment activities that allow a user (or a computer) to send an inquiry (or a file) from a specified terminal to a specified computer and receive a valid response within a specified amount of time. Unless someone is able to facilitate the typing of something on a terminal and the receipt of a response, then he or she is not a datacom specialist but, at best, only supports a subfunction. Often telecommunications professionals fail to appreciate what data communications really is, and simply think that providing a digital pipe is the end of the story.

An obvious approach to the challenge of providing true connectivity between devices is to deploy networking software for common hardware and a single operating system, as described earlier. Such vendor-proprietary or vendor-dominated solutions were the hallmark of the 1970s and 1980s and have been discounted by the user community, which votes with its purse. As more and more organizations are linked, however, a single-vendor environment quickly becomes unrealistic. A standards-based approach is needed.

The Open Systems Interconnection Basic Reference Model and related standards (see Sec. 2.4) provide interoperability at the communication level, when (1) the standards are implemented, (2) the same suite of standards is used, and (3) vendors' agreements have been reached to select a common parameter profile. IEEE Portable Operating System Interface (POSIX) standards facilitate multivendor operation by addressing the interface of the communication fabric with application programming.[39] DCE specifies programming interfaces and protocols for network computing in a suitable form for standardization. This tackles the limitations of existing systems. For example, the Open Systems Interconnection suite and the Internet suites do not directly incorporate security or provide application program support; DECnet and Apollo's Domain do not provide multivendor support directly; NSF does not support WAN network computing or security directly; NetWare does not support WAN network computing directly.

Many computer vendors, representing the majority of the market in PCs and workstations, are now in the process of porting DCE software to their operating systems. End users will be able to buy DCE implementations as a part of the operating system software on platforms employing, among others, Microsoft Windows, IBM's MVS, Unix, and VMS operating systems. DCE software is installed on each computer in a system and accesses the network through a transport-layer interface such as TCP/IP or ISO's Transport Protocol.

An application that runs on a multivendor network must meet the programming requirements of that network. These include synchronizing with programs on different systems, buffering large transfers, converting data presentations when necessary, and verifying security privileges.[37] DCE software automates the process, making it practical to create reliable, secure applications running cooperatively on a mixture of supporting hardware and OS platforms. There have been efforts to standardize APIs with an approach similar to the one shown in Fig. 2.27. DCE goes even beyond that by supporting security and real-time cooperative computing.

The core set of services for DCE includes an RPC technology, naming and X.500 directory services, Kerberos (MIT), distributed file services with POSIX 1003.1 semantics, a threads package with the POSIX 1003.4a interface, time services to synchronize computer clocks with the universal time coordinate, and PC file services.[30]

In network computing applications, a task may be initiated on one system and executed transparently on another system, which may

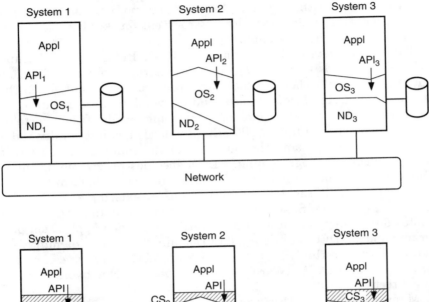

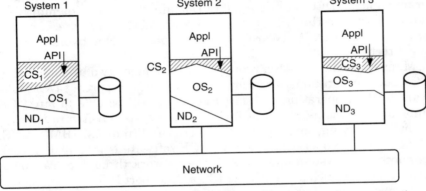

Figure 2.27 Eliminating some incompatibilities to support network computing (common APIs). OS = operating system. ND = network driver. CS = compatibility software (details are OS-dependent, but provide a common mechanism).

have a completely different architecture. The client computer initiates an RPC at the application level. The next layer encodes the procedure's parameters and the data associated with them in an architecture-independent way. This information is placed into buffers. The information is transferred to the communication server by a run-time library of routines using an appropriate transport protocol. The server then transfers the information over the physical network to the remote server. The remote server decodes the buffer and delivers the data to the remote application program for processing (see Fig. 2.28). The results are returned to the client using a similar process. All of this is completely transparent to the user; the task could have been

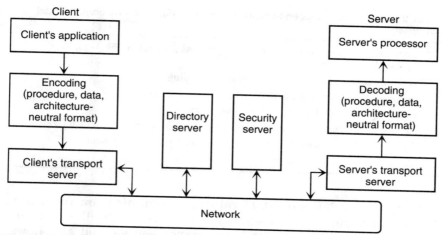

Figure 2.28 Distributed computing environment.

accomplished in the user's workstation or thousands of miles away without any discernible difference. A DCE directory service allows a client to determine the address of the required database server. When a client logs into the DCE infrastructure, a security server authenticates the user. As RPCs are issued by the client, the requests are automatically routed to the security server, which determines (using access control lists) if this client can access the specified target server.

DCE utilizes the concept of cells (not to be confused with an ATM cell). A cell is a collection of computers under a common administration. It might consist of a few or few thousand computers. A corporation could use one cell per organization. Cell names and locations are registered in a worldwide directory. Each cell maintains its own directory, security, and file services on centralized machines.[37] Communication takes place intracell and intercell. DCE clearly relies heavily on communication services.

2.4 Open Systems Interconnection Basic Reference Model

The description of LAN, frame relay, SMDS, and ATM protocols in the following chapters will follow the terminology of the Open Systems Interconnection Basic Reference Model (OSIBRM), which has been available since 1984. A short description of this communications model is provided below. To facilitate network interconnection, standards for open systems have been developed by the International Organization for Standardization (ISO).

TABLE 2.9   Layers of the Open Systems Interconnection Basic Reference Model (OSIBRM)

7	Application	Support of user functions such as association, file transfer, directory, etc.
6	Presentation	Transfer syntaxes (character coding)
5	Session	Coordination functions, synchronization
4	Transport	Reliable end-to-end communication
3	Network	Delivery within a single subnetwork, addressing, multiplexing
2	Data link	Delivery of frames (blocks) of data between two points
1	Physical	Bit transmission

A *layer* is a defined set of related communication functions. Protocols describe ways in which remote peers can utilize functions within a layer. Seven major layers have been defined as follows: application (7), presentation (6), session (5), transport (4), network (3), data link (2), and physical (1) layers. Table 2.9 identifies key functions. This model is described in the specification ISO 7498 and also in CCITT X.200. The term *upper layers* (or *higher layers*) refers to layers 4 to 7; *lower layers* include layers 3 to 1.

The higher adjacent layer is called the *user,* the lower one is called the *provider* (the term here does not refer to the ultimate end user). User and provider describe, respectively, the relationship between the consumer and the producer of a layer service. As one moves through layers, users become providers and vice versa. Communication with a remote peer, at the same layer, involves a protocol, as depicted in Fig. 2.29. Adjacent entities communicate by exchanging primitives with each other via the *service access point* (SAP). The SAP is a conceptual delivery point, and as such it can be addressed.

The upper layers—application, presentation, session, and transport—are generally, although not always, independent of the telecommunications network; the reason for the exception is that some carriers may offer functionality above the network layer, for example, e-mail. In general, however, these layers are components of the end-user systems and are insulated from networking operations.

A *primitive* represents the logical exchange of information between a layer and the adjacent layers; it does not specify or constrain implementation. Two kinds of service are available: confirmed and unconfirmed. A confirmed service produces information from the remote peer entity on the outcome of the service request (this may be needed when additional action is contingent upon a successful outcome). An unconfirmed service only passes a request along; this is a faster interaction, since there is no overhead involved with the response.

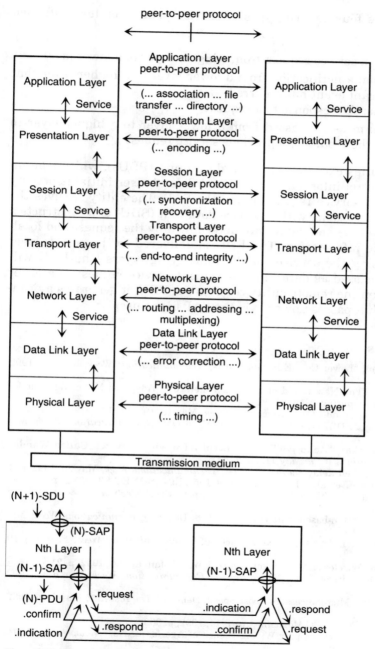

Figure 2.29 Peer-to-peer protocols.

There are four generic types of service primitives for confirmed service:

1. Request: a service request from a higher layer to a lower layer
2. Indication: a notification from a lower layer to a higher layer that a significant event has occurred
3. Response: the response to a request
4. Confirm: message passed from a lower layer to a higher layer to indicate the results of a previous service request

Peer entities exchange protocol data units (PDUs), which contain protocol control information (PCI) and data. A user initiates activity by issuing a service request across the SAP. The entity receives the service request along with a service data unit (SDU) and constructs a PDU whose type and values are determined by the request and locally available information. The PDU is delivered to the remote peer partner using the services of the underlying layers (the PDU will itself be enclosed as data in a subsequent service request to a lower layer). When the remote entity receives the PDU, it generates a primitive, which it passes upward via the SAP to the user.

References

1. M. Semilof, "Lower-Cost Ethernet Cards," *Communications Week,* May 11, 1992, p. 8.
2. S. Morse, "The New Breed of 16-Mb/s Token-Ring Adapters Make Big Strides," *Network Computing,* March 1992, pp. 20 ff.
3. LAN Wiring Trends, *Communications Week,* May 18, 1992, p. 1.
4. M. Semilof, "FDDI-Over-Copper from IBM, 3COM," *Communications Week,* May 25, 1992, p. 5.
5. S. Girishankar, "Adaptive Targets Local ATM with a Switch Called Wanda," *Communications Week,* May 11, 1992, p. 1.
6. R. Gareiss, "Cell Relay Service Planned," *Communications Week,* May 25, 1992, p. 1.
7. T. C. Banwell et al., "Transmission of 155 Mb/s (SONET STS-3) Signals over Unshielded and Shielded Twisted-pair Copper Wire," *Electronics Letters,* vol. 28, no. 2, 1992, p. 1.
8. M. Semilof, "Hub-based Wireless Network to Debut," *Communications Week,* May 11, 1992, p. 2.
9. M. Dortch, "Wireless LANs: Now Practical?," *Communications Week,* February 10, 1992, p. 11.
10. Worldwide Wireless LAN Market, *Network World,* January 20, 1992, p. 21.
11. M. Dortch, "Wireless LANs Move Ahead," *Communications Week,* March 23, 1992, p. 14.
12. IEEE, "RISC Microprocessors Boost Speed, Density," *The Institute,* May/June 1992, p. 1.
13. J. Prosise, "Tutor," *PC Magazine,* April 14, 1992, p. 353.
14. B. Bhushan, "Twisted-pair Wiring Comes of Age for LANs," *Networks In-Depth,* July 1991, pp. 1 ff.
15. Digital Equipment Corporation, *A Primer to FDDI,* EC-H0750-42 LKG, 1991.
16. E. Mier, "Hubs: An Embarrassment of Riches," *Communications Week,* May 25, 1992, pp. 49 ff.

17. J. Cummings, "Hub Vendors Vary in Market Strategies," *Network World*, May 11, 1992, p. 17.
18. J. S. Skorupa, "Evaluation of Intelligent Wiring," *COMNET 92*, Washington, D.C., January 1992.
19. S. Girishankar, "Cameo to Offer Low-Cost Managed Ethernet Hub," *Communications Week*, March 9, 1992, p. 5.
20. J. Mulqueen, "HP Cuts Managed-Hub Prices," *Communications Week*, February 3, 1992, p. 5.
21. D. Berge, "10Base-T Hub Price Wars," *Communications Week*, December 21, 1992, p. 15.
22. S. Girishankar, "Applications to Land on Hubs," *Communications Week*, May 11, 1992, p. 1.
23. D. Minoli, *Telecommunications Technology Handbook*, Artech House, Norwood, Mass., 1991.
24. D. Minoli,"Internetworking LANs: Repeaters, Bridges, Routers & Gateways," *Network Computing*, October 1990, pp. 96 ff.
25. P. Cope, "New LAN Bridges Adapt to Changing User Needs," *Network World*, May 11, 1992, p. 5.
26. Data Communication, "Sharing the Load: Client/Server Computing," March 21, 1989, pp. 19–29.
27. L. Berg, "Implementing Client/Server Computing," *COMNET 92*, Washington, D.C., January 1992.
28. K. Myhre, "Please Explain Client/Server," *COMNET 92*, Washington, D.C., January 1992.
29. S. Semilof, "Users Unsure of Need for LAN Superservers," *Communications Week*, February 24, 1992, p. 1.
30. D. Ferris, "Client/Server Database Models Are Emerging," *Network World*, May 11, 1992, p. 19.
31. A. Dickman, Bellcore, personal communication, June 1992.
32. D. Minoli, "APPN or APPI?," *Network Computing*, February 1993, p. 126.
33. P. Maclean, "Peer-to-Peer LANs," *Network Computing*, March 1992, pp. 85 ff.
34. McData Corporation promotional literature, Broomfield, Colo.
35. C. Partridge, *A Proposed Flow Specification*, Internet Research Task Force, Request for Comment series, 1992.
36. *Frame Relay vs. SMDS vs. T1*, Probe Research Report, Ceder Knoll, N.J., 1992.
37. D. Hartman, "Unclogging Distributed Computing," *IEEE Spectrum*, May 1992, pp. 36 ff.
38. D. Minoli, *Enterprise Networking, Fractional T1 to SONET, Frame Relay to B-ISDN*, Artech House, Norwood, Mass., 1993.
39. ISO/IEC, Information Technology—Portable Operating System Interface (POSIX), Part 1: System Application Interface (API) [C Language], IEEE/ANSI 1003.1-1990; ISO 9945-1, 1990.

3

LAN Basics:
First-Generation
Lower Layers

3.1 LAN Technology

This chapter examines key features of first-generation LANs, including medium-sharing methods, lower-layer protocols, upper-layer protocols, and addressing.

3.1.1 LAN topologies

There are three major physical LAN topologies (see Fig. 3.1): star, ring, and bus. A *star* network is joined at a single point, generally with central control (such as a wiring hub). In a *ring* network the nodes are linked into a continuous circle on a common cable, and signals are passed unidirectionally around the circle from node to node, with signal regeneration at each node. A ring with a central control is known as a loop. A *bus* network is a single line of cable to which the nodes are connected directly by taps. It is normally employed with distributed control, but it can also be based on central control. Unlike the ring, however, a bus is passive, which means that the signals are not regenerated and retransmitted at each node.

Other configuration variations are available, particularly when looking at the LAN from a strictly physical perspective: the *star-shaped ring* and the *star-shaped bus*. The first variation represents a wiring methodology to facilitate physical management: at the logical level the network is a ring; at the physical level it is a star centralized at some convenient point. Similarly, the second variation provides a

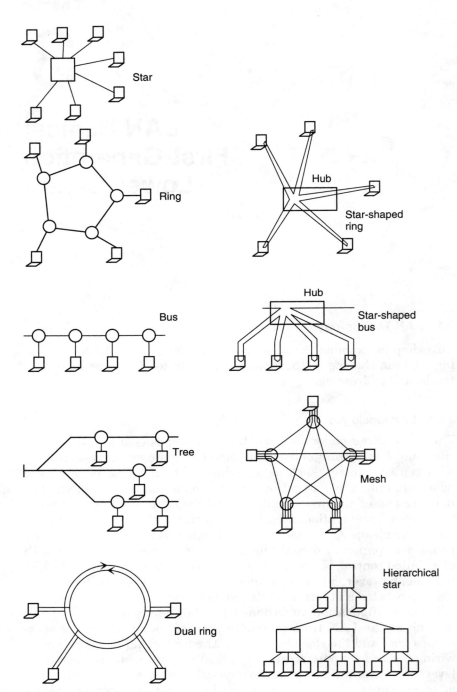

Figure 3.1 Network topologies.

TABLE 3.1   LAN Topologies

LAN	Early	Recent
First generation, broadband	Bus, tree	Bus, tree
First generation, Ethernet	Bus	Star-shaped bus
First generation, token ring	Ring	Star-shaped ring
Second generation	Fiber double ring	Star-shaped double ring
Third generation	Star-based access segments	—

logical bus, but wired in a star configuration using wiring hubs. The *hierarchical star* is a common wiring approach utilized in actual LAN wiring arrangements. It uses a set of cascaded hubs to build a hierarchical network based on some local policy (e.g., one hub per department, floor, or workgroup).

Other topologies such as the *tree,* the *mesh,* and the *double ring* have been used to various degrees. The tree has been used in CATV applications and may be used in some analog broadband LANs. The mesh is more typical of WANs, but could also be used in some LAN applications. The double ring provides two rings: one for transmission and one for reception. This method is used, for example, in FDDI.

Table 3.1 summarizes the use of these topologies in the three generations of LANs.

3.1.2   LAN bandwidth

The *nominal capacity* of a communication link depends on (1) the medium used to transmit the information (e.g., twisted-pair wire, coaxial cable, optical fiber); (2) the length of the path; and (3) the encoding scheme. Coaxial cable has a bandwidth of 300 to 400 MHz. Twisted-pair cable (of standard gauge) has a bandwidth of 1 to 5 MHz at local loop lengths and 20 to 40 MHz up to a few meters. Optical fibers have bandwidths in the 2 to 6 GHz region. First-generation LANs generally used coaxial cable; more recently, twisted-pair wire systems have appeared. Encoding refers to physical layer mechanisms to create signals representing the bits that can be carried by the medium at hand. Simple mechanisms encode 1 bit per baud; more sophisticated multilevel mechanisms encode multiple bits per baud. Multilevel methods support higher digital throughput and are used, for example, in the twisted-pair LANs and FDDI over twisted-pair medium.

Bandwidth over a medium can be allocated using two traditional techniques among others: frequency division multiplexing (FDM) and time division multiplexing (TDM). FDM divides the available bandwidth into separate channels of appropriate bandwidth. TDM starts with a specified (digital) bandwidth (which depends on the three factors just described) and allows several nodes to access the channel on a rotating deterministic basis. Each node in turn is given access to the channel for long enough to transmit part of its information; the information is reassembled at the destination. In traditional LAN terminology, a system based on FDM is known as *broadband*. A LAN in which the signals are put on to the medium at their original frequency is called *baseband*; and one in which all the signals are modulated on to a single carrier frequency is called *carrierband*. In baseband transmission, the unmodulated data are pulsed in a single channel directly onto the transmission medium. Baseband LANs do not employ FDM to derive multiple channels from the medium, but employ distributed random-access techniques somewhat similar to statistical TDM to achieve medium sharing. These techniques are known as medium access control.

The advantage of a traditional broadband LAN is its multichannel capacity; however, it is more expensive than a carrierband system. The thrust of third-generation LANs is to support high-capacity isochronous traffic (such as that generated by voice and video applications) without having to use analog FDM techniques.

The actual capacity of a LAN depends on details of the encoding mechanism as well as the entire protocol suite. For example, FDDI systems use a method that sends 5 bits every 4 bits of user data (this is called 4B/5B encoding); other systems use 8B/10B methods (see Chap. 9). However, the most critical factors affecting the actual throughput are the upper-layer protocols, starting with the MAC. In its effort to support a distributed medium-sharing discipline, approximately 20 percent of the bandwidth is used up in collisions (in Ethernet—with a similar overhead for token systems). The upper layers (data link layer, network layer, transport layer, etc.) all add overhead, thereby limiting the actual efficiency of the transmission. Segmentation, framing, and cellularization also add overhead. Additionally, there can be retransmission delays as well as other processing delays, all impacting the actual throughput. These delays, the reduction in throughput, and the variance of the interframe delay become critical when transporting isochronous traffic.

3.1.3 Medium-sharing disciplines

There are two common ways of ensuring that nodes gain orderly access to the network, and that no more than one node at a time gains control of the shared LAN channel. (There are many other such techniques, but these are not used in a cabled-LAN context.) The first is by contention method; the second is by a variant of polling.

Carrier sense multiple access with collision detection (CSMA/CD). If node A has a message to send, it checks the shared-medium network until it senses that it is traffic-free ("listen-before-talk"), and then it transmits. However, since all the nodes on the network have the right to contend for access (hence "multiple access"), node A keeps monitoring the network in case a competing signal has been transmitted simultaneously with its own ("listen-while-talk"). If a second node is indeed transmitting, the two signals will collide. Both nodes detect the collision by a change in the channel energy level, stop transmitting, and wait for a random time (of the order of microseconds) before attempting to regain access.

Contention techniques are nondeterministic; they cannot guarantee access to the network within a specified amount of time, although in a well-designed network, the probability that this occurs can be made high.

Token. Token-based LANs avoid the collisions inherent in Ethernet by requiring each node to defer transmission until it receives a "clear-to-send" message, called a token. The token is a special control packet which is circulated around the network from node to node, in a preestablished sequence, when no transmission is otherwise taking place. The token signifies exclusive right to transmission, and no node can send data without it. Each node constantly monitors the network to detect any packet addressed to it; this could be a message or the token (the token does not have an address). When the token is received by a node, and the node has nothing to send, then the node passes it along (without any channel access delay) to the next node in the sequence. If the token is accepted, it is passed on after the node has completed transmitting the data it has in its buffer. The token must be surrendered to the successor node within a specific time, so no node can monopolize the network resources. Each node knows the address of the predecessor and the successor. The sequence by which the token is circulated to the nodes is set by network management and does not have to necessarily conform to the physical bus or ring location.

Token technology has been adopted by large vendors, such as IBM, while the Ethernet technology has been brought to the market by DEC and many smaller vendors.

It is worth noting that as developers sought to increase the throughput from 10 Mb/s to 100 Mb/s in FDDI, the collision-based Ethernet access method was found unsuitable, and the more "orderly" token method was employed. By appropriately setting the token holding time, stations can, in effect, be provided with a more consistent throughput than possible with Ethernet. In evolving beyond 100 Mb/s total (i.e., aggregated *across all users*) to 155 Mb/s and/or 622 Mb/s *per user,* developers are finding that noncontention ATM techniques are much better suited.

3.2   Lower-Layer LAN Protocols

In the early 1970s, the Xerox Palo Alto Research Center began an exploration of packet switching on coaxial cable which ultimately led to the development of Ethernet LANs. The original Ethernet standard was issued in 1978 by Digital Equipment Corporation, Intel Corporation, and Xerox Corporation. This standard is also known as *DIX Ethernet* (where the letters stand for the companies). Eventually, formal standards became available.

3.2.1   Standards

In a LAN environment, layer 1 and 2 functions of the OSI reference model have been defined by (1) the IEEE 802 standards for first-generation LANs, (2) ANSI X3T9.5 for second-generation LANs, and (3) industry groups such as the ATM Forum (particularly the Local ATM group), ECSA T1S1, and CCITT* (the last two bodies having standardized the supporting ATM functions) for third-generation LANs. Using "internetworking" protocols defined at layer 3 (such as IP) and, typically, connection-oriented transport protocols (such as TCP), one can then build the protocol suite up to layer 7 in order to support functions like e-mail, file transfer, directory, etc.

The use of *transmission control protocol / internet protocol* (TCP/IP) has been commercially common for the *upper layers,* particularly for LANs interconnected via the Internet. TCP/IP is a well-established standard, discussed later: TCP/IP-related products constitute a $7 billion per year business. Internet consists of over 2000 networks with more than 180,000 hosts and 1 million users; also, there are over 2000 large private networks, not on Internet, which use TCP/IP. Some LANs use vendor-proprietary upper layers, for example, Apple with AppleTalk. An OSI-based upper-layer suite (from the network layer upward) is also possible. The availability of international OSI standards may exert a force toward possible migration to these new standards, although many are skeptical about such prospects.

Below is a description of the LAN standards at layers 1 and 2. Table 3.2 provides a list of the key IEEE and ISO standards. Section 3.3 describes TCP/IP standards.

The charter of IEEE Project 802 encompassed the physical and data link layers (see Fig. 3.2). The range of transmission speed was set at 1 to 10 Mb/s. Because LANs are based on a shared medium, the link layer had to be split into two sublayers. These sublayers are the medium access control, and the logical link control (LLC). The LLC sublayer provides a medium-independent interface to higher layers.

---

*Known as Telecommunication Standardization since 1993.

TABLE 3.2   Recently Published IEEE and ISO LAN Standards

IEEE 802-1990	Overview and Architecture.
IEEE 802.1D-1990	Medium Access Control Bridges.
IEEE 802.1E-1990	System Load Protocol.
IEEE 802.3h-1990	Supplement to Carrier Sense Multiple Access with Collision Detection: Access Method and Physical Layer Specifications: Layer Management.
IEEE 802.3h-1990	Supplement to Carrier Sense Multiple Access with Collision Detection: Access Method and Physical Layer Specifications: System Considerations for Multisegment 10 Mb/s Baseband Networks and Twisted-Pair Medium Attachment Unit and Baseband Medium, Type 10Base-T.
IEEE 802.5-1989	Token Ring Access Method and Physical Layer Specifications.
IEEE 802.5b-1991	Recommended Practice for Use of Unshielded Twisted-Pair Cable for Token Ring Data.
IEEE 802.5c-1991	Supplement to Token Ring Access Method and Physical Layer Specifications: Recommended Practice for Dual Ring Operation With Wrapback Reconfiguration.
IEEE 802.6-1990	Distributed Queue Dual Bus Subnetwork of a Metropolitan Area Network.
IEEE 802.7-1989	Recommended Practice for Broadband Local Area Networks.
ISO/IEC 8802-2:1989	Information Processing Systems, Local Area Network, Part 2: Logical Link Control.
ISO/IEC 8802-3:1992	Information Technology, Local and Metropolitan Area Networks, Part 3: Carrier Sense Multiple Access with Collision Detection: Access Method and Physical Layer Specifications. (This standard supersedes both *ISO/IEC 8802-3:1990* and *IEEE 802.3b, c, d, e-1989*. It contains Broadband Medium Attachment Unit and Broadband Medium Specifications, Type 10Broad-36; and Physical Signaling, Medium Attachment, and Baseband Medium Specification, Type 1Base-5.)
ISO/IEC 8802-4:1990	Information Processing Systems, Local Area Network, Part 4: Token-Passing Bus Access Method and Physical Layer Specifications. (This standard supersedes *IEEE 802.4-1985*.)

The MAC procedure is part of the protocol that governs access to the transmission medium. This is done independent of the physical characteristics of the medium, but taking into account the topological aspects of the subnetwork. Different IEEE 802 MAC standards represent different protocols used for sharing the medium. Contention access uses carrier sense multiple access with collision detection (CSMA/CD) techniques and is representative of the Ethernet bus architecture (IEEE 802.3) (see Table 3.3).

Noncontention methods use tokens, typically in a ring configuration. IBM introduced their proprietary LAN in 1984 by way of their 4-Mb/s IBM Token Ring System (at about the same time IBM introduced a proprietary wiring scheme, the IBM Cabling System, which

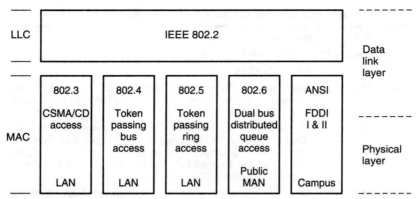

Figure 3.2 First- and second-generation LAN standards.

TABLE 3.3 Functions at Specified Protocol Levels

LLC	Reliable transfer of frames; connection to higher layers
MAC	Addressing; frame construction; token and collision handling
PHY (physical layer protocol: explicit only in more recent standards such as FDDI, SONET, LATM)	Encoding and decoding; clocking
PMD (physical medium dependent: explicit only in more recent standards such as FDDI, SONET, LATM)	Cable parameters (optical and electrical); connectors

utilized shielded twisted-pair). Token technology has now been embodied in the IEEE 802.5 standard. Token bus systems are described in IEEE 802.4. Token ring systems also now operate at 16 Mb/s, although no IEEE standard exists at this speed. While the token ring technology has been widely deployed in the business community, Ethernet still has a major market presence, particularly in the scientific, academic, and manufacturing environments.

3.2.2 Logical link control

IEEE Std 802.2-1985 (ISO 8802-2:1989) provides a description of the peer-to-peer protocol procedures that are defined for the transfer of information and control between any pair of data link layer entities on the LAN.

The LLC procedure is that part of the protocol that specifies the assembling of data link layer frames and their exchange between data stations, independently of how the transmission medium is shared. The LLC sublayer supports medium-independent data link

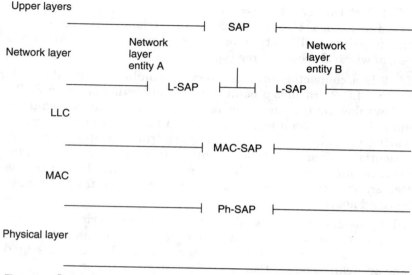

Figure 3.3 Lower-layer LAN service access points and protocols.

functions and employs the MAC sublayer service to provide services to the network layer. The protocol is important because it provides a uniform interface between higher layers and the MAC protocols of 802.3, 802.4, and 802.5, as well as IEEE 802.6 and FDDI. Effectively, it provides transparency to the network layer with respect to the underlying LAN medium and, thus, transparency to application software (e-mail, word processing, etc.).

A number of different SAPs are defined in the IEEE 802 protocol, as seen in Fig. 3.3 (refer to the end of Chap. 2 for a refresher). The SAP can be considered an address within a station that identifies a particular application or service. Note that there are multiple LLC SAPs (L-SAPs). Service primitives are used to exchange information between the different layers. This exchange of information embodies the service provided by these sublayers. LLC connects two peer L-SAPs in the two end systems (LAN stations). Three frame formats are defined for LLC: information transfer frames, supervisory frames, and unnumbered frames. As discussed next, there are three protocol types: LLC 1, LLC 2, and LLC 3. The specific use of these frames depends of the type of LLC operation utilized.

- *LLC 1* provides an unacknowledged connectionless data link service. It allows sending and receiving of frames between SAPs without the prior establishment of a connection between the two communicating endpoints; no call setup or call termination phase is required. No guarantee of delivery or sequentiality is provided by

LLC 1, but this can be accomplished at the transport layer, if desired or needed. Unacknowledged connectionless service may be point-to-point, multicast, or broadcast. Unnumbered frames are used, of which there are three types.

- *LLC 2* is a connection-oriented service, similar to virtual circuit service. LLC 2 provides a point-to-point connection between SAPs; it allows flow control and error recovery. A call setup procedure is required to establish a logical connection between the two communicating endpoints prior to exchanging frames containing data. Sequential delivery of frames is guaranteed: data frames contain sequence numbers and the frames must be acknowledged by the receiver. This type of operation requires all three frame formats identified above.

- *LLC 3* provides an acknowledged connectionless service. Here user data is sent in an Acknowledged Connectionless command frame (a newly defined unnumbered frame) and must be acknowledged using an Acknowledged Connectionless response frame.

LLC defines two sets of primitives, in order to fulfill its user and provider functions. Primitives between the LLC sublayer and the higher layers are prefixed by "L"; primitives between the LLC and the MAC sublayers are prefixed by "MA."

Unacknowledged connectionless LLC 1 service supports only the higher layer-to-LLC sublayer L-SAP primitives necessary for one instance of data transmission, without prior connection establishment. Hence, the service primitives are simply *LDATA.request* and *LDATA.indication.* Connection-oriented LLC 2 requires several types of higher layer-to-LLC sublayer L-SAP primitives, to undertake such activities as logical link establishment and deestablishment, flow control, and so on.

The MAC-SAP primitives between the LLC sublayer and the MAC sublayer provide for the transfer of data between the two sublayers. The interface and primitives are independent of the MAC specifics (whether 802.3, 802.4, or 802.5, described in the next section).

Figure 3.4 depicts the format of an LLC frame, which, in turn, is enveloped inside the information field of the lower MAC sublayer frame, as shown in the second part of the figure. The LLC standard includes both the destination and source address fields within the header. These addresses are associated with the SAP of the LLC sublayer, and their values are defined to be unique only within a given station address (so that after a frame has been routed to the appropriate station, it undergoes additional internal routing to reach the proper network layer entity). A network layer entity, which is a user of LLC services, can be reached *through an address formed by the*

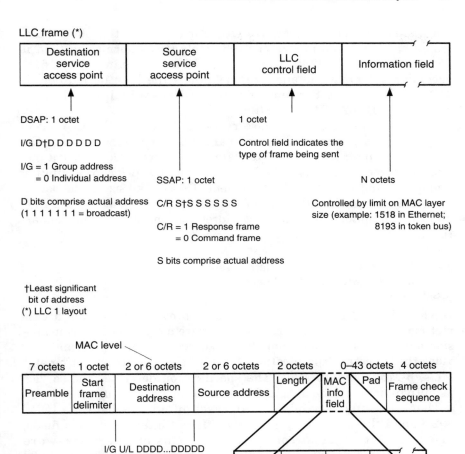

Figure 3.4 LLC and IEEE 802.3 (ISO 8802-3) MAC frame structure.

concatenation of the L-SAP with a given (physical) station address. The station address identifies the service access point associated with the MAC entity within each system. This address has a one-to-one mapping with the physical address or physical SAP. (The physical address is not explicitly carried in each frame, but is implied by the physical connection of the station to the network.) Addressing is revisited later.

The maximum size of the LLC data field is determined by the maximum frame size imposed by the MAC layer, since the LLC PDU is encapsulated inside the MAC PDU. For example, the 802.4 MAC frame has a maximum size of 8193 octets (the information field is 8174 octets) and 802.3/Ethernet MAC has a maximum of 1518, excluding the preamble and start delimiter (the information field is 1500 octets or less). For comparison, FDDI has a limit around 4500 and the IEEE 802.6 MAN at 9188.

A minimum frame size is also required for correct MAC protocol operation and is specified by the particular implementation of the standard. If necessary, the MAC data field is extended by extra pad bits in units of octets after the LLC data field but prior to calculating and appending the frame check sequence (FCS). The size of the pad, if any, is determined by the size of the data field supplied by the LLC and the minimum frame size and address size parameters of the particular MAC implementation.

3.2.3   Medium access control: Carrier sense multiple access with collision detection (CSMA/CD)

IEEE Std 802.3-1985 (ISO 8802-3:1989) provides a medium access method by which two or more stations share a common-bus transmission medium. The standard applies to several medium types and provides the necessary specifications for a baseband LAN operating at 1 Mb/s, 5 Mb/s, and 10 Mb/s. The specification describes the service primitives between the MAC sublayer and the physical layer.

There are only very minor MAC-level differences between IEEE 802.3 and Ethernet systems (slightly different interpretation of fields, discussed later). Most commercial systems on the market are Ethernet-based. People commonly refer to the technology with the nomenclature IEEE 802.3/Ethernet.

The 802.3 standard defines a "logical bus"; many implementations, however, have physical configurations that topologically may not be a bus, as described at the beginning of this chapter. Ethernet, and the initial 802.3 standard based on it, operate at 10 Mb/s over a coaxial-cable bus, with all stations connected to one transmission line. The IEEE 802.3 group has subsequently built on its original effort by developing new standards, particularly with reference to twisted-pair LANs. The four variants of the 802.3 LANs are as follows (see Table 3.4):

- *10Base-T.*   Baseband LAN operating at 10 Mb/s over twisted-pair, allowing a distance between station and hub of 100 m. Topologically, this LAN has a star configuration.

- *1Base-5.*   Baseband LAN operating at 1 Mb/s over twisted-pair, allowing a distance between station and hub of 250 m. Topologically, this LAN utilizes a star configuration.

TABLE 3.4   Topological Parameters

	Maximum distance per segment	Connections per segment
Twisted-pair	100 m	232
10Base-5	500 m	100
10Base-2	185 m	30

- *10Base-5.*   Baseband LAN operating at 10 Mb/s over coaxial cable, with a maximum bus length of 500 m. This LAN is the closest to the original Ethernet. The standard was approved in 1983. In the original 802.3 standard a station or server is connected with the coaxial bus using heavy-gauge shielded and twisted copper wires. Shielding and pairing is employed to minimize electromagnetic and radio frequency interference.

- *10Base-2.*   Baseband LAN operating at 10 Mb/s over thin coaxial cable (type RG-58) with a maximum bus length of 185 m. Regular coaxial cable is approximately 1 cm thick; this thickness makes it difficult to bend the cable in tight office environments. Thin cable performs almost as well as regular coaxial, except that signals attenuate more rapidly.

Twisted-pair installations accounted for 57 percent of all LANs deployed, and by press time it was estimated that over 80 percent of all new LAN shipments will use twisted-pair media.[1]

10Base-T.   Ethernet was initially specified to operate over "thick" coaxial cable. The wiring topology was a bus; the nodes hang off the bus at predefined distances (a minimum of 2.5 m). Because of the expense of the cable and the fact that requirements for broadband did not immediately arise, a cheaper and thinner coaxial cable was developed (this is also known as "cheaper-net" or "thin-net"). This daisy-chain configuration only supports baseband applications, but it does support the 10 Mb/s nominal throughput. In addition, this cable is easier to install than coaxial. However, this cable is still more expensive than twisted-pair cable.

The 10Base-T IEEE specification allows Ethernet LAN users to utilize the existing telephone twisted-pair cable (typically, 24 American wire gauge solid copper wire with unshielded PVC insulation, and containing four or more pairs) as the physical medium for 10-Mb/s transmission over segments up to 100 m. The 10Base-T Ethernet requires two pairs of wire to each service location; since most telephones operate on a single pair and most modern buildings in the United States are cabled with four-pair cable, usually there are spare pairs in place at each work area. The 10Base-T medium access unit must coexist with other signals that may be present on the other pairs in the telephone cable (including 1Base-5 LANs, token-passing LANs, analog voice, and ISDN). In addition to specifying the requirements for the transmitter, the receiver, and the attachment unit interface electronics contained within a 10Base-T medium access unit, the specification also places performance constraints on the link segment between the access units. (These specifications are critical to the proper functioning of the Ethernet LANs built on this type of cables.)[2]

It has backward compatibility with existing Ethernet hardware at the attachment unit interface (the standard 15-pin D connector). Users can replace the coaxial-based medium access units with the 10Base-T medium access units, protecting existing investment in Ethernet controllers (see Fig. 3.5). The 10Base-T twisted-pair Ethernet standard received final approval in 1990. Twisted-pair Ethernet products were popular even before the standards were finalized because of their cheaper, and often already-installed, wire.

Comparison of Ethernet and IEEE 802.3.   The IEEE 802.3 standard (and the international document ISO 8802-3:1989), is a more recently defined standard than the original DIX Ethernet. The difference between the two standards is in the use of the fifth header field, which contains a protocol-type in an Ethernet frame, and the length of the data field in the "IEEE 802.3" frame, as seen in Fig. 3.4.

The protocol-type field in Ethernet is used to distinguish between different protocols operating on the coaxial cable and facilitates their coexistence on the same physical cable. The maximum length of an Ethernet frame is 1526 octets (1519 for the actual frame and 7 octets of preamble), with a data-field length up to 1500 octets. The length of the 802.3 data field is also 1500 octets for 10 Mb/s LANs, but it is different for other transmission speeds. The actual length of the data field is therefore indicated in the 802.3 header; the type of protocol it carries is indicated here in the 802.2 LLC header contained in the data field of the 802.3 field depicted in Fig. 3.4.

Both frame formats can coexist on the same physical cable. This is done by using protocol type numbers (type field) greater than 1500 in the Ethernet frame. (However, different PC drivers are needed to handle each of these formats.) This implies that Ethernet MAC and the IEEE 802.3 MAC are compatible. However, the Ethernet LLC and the IEEE 802.3/802.2 LLC are not compatible.[3]

The 802.2 LLC layer above IEEE 802.3 MAC uses the L-SAP, which has a 3-byte header comprised of an 8-bit destination service access point (DSAP) field, an 8-bit source service access point (SSAP) field, and an 8-bit control field, as shown in Fig. 3.4. Numbers for these fields are assigned by an IEEE committee (see Table 3.5).

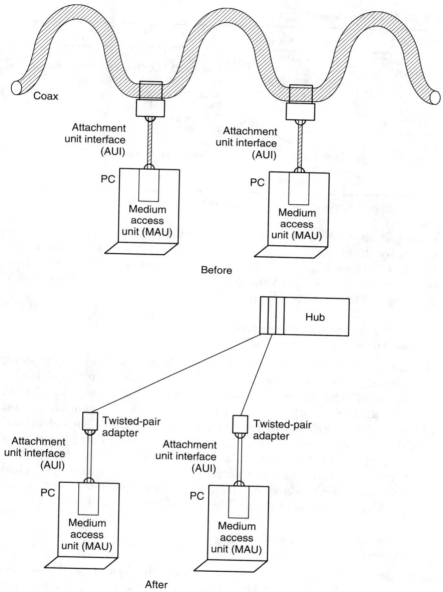

Figure 3.5   Retention of MAU with 10BASE-T LANs.

In 1986 an extension was made to the IEEE 802.2 LLC Type 1 protocol, to support the SubNetwork Access Protocol (SNAP). It is an extension to the L-SAP header just described. SNAP use is indicated by the value 170 in both the SSAP and DSAP fields in the LLC frame shown in Fig. 3.4. The SNAP header consists of 3 octets showing an

TABLE 3.5   L-SAP Values

L-SAP*	Internet notation	Description
00000000	0	Null SAP
01000000	2	Individual LLC Sublayer Management
11000000	3	Group LLC Sublayer Management
00100000	4	SNA Path Control
01100000	6	DoD Internet Protocol
01110000	14	Proway LAN
01110010	78	EIA RS 511
01110001	142	Proway LAN
01010101	170	SubNetwork Access Protocol (SNAP)
01111111	254	ISO 8473
11111111	255	Global DSAP

*Numbers in binary form, with most significant bit shown at left.

organization code, known as "organizationally unique identifier," followed by 2 octets showing an "EtherType" field (also known as Protocol ID) (see Fig. 3.6). This approach allows IEEE 802.2 encapsulation on 802.3, 802.4, and 802.5 LAN data using the SNAP to specify the type of Ethernet being employed.

3.2.4   Medium access control: Token-passing bus access method

IEEE Std 802.4-1985 (ISO 8802-4:1989) describes the token-passing bus access method and its associated physical signaling and media technologies. The access method coordinates the use of the shared medium among the attached stations. It specifies the electrical and physical characteristics of the transmission medium, the electrical signaling used; the frame formats of the transmitted data, the actions of a station upon receipt of a data frame, and the services provided at the SAP between the medium access control sublayer and the logical link control sublayer above it. The specification also describes the service primitives between the MAC layer and the physical layer.

3.2.5   Medium access control: Token-passing ring access method

IEEE Std 802.5-1985 (ISO 8802-5) specifies the formats and protocols used by the token-passing ring system at the control sublayer and physical layer. It also specifies the means of attachment to the token-passing ring access method. The protocol defines the frame format, including delimiters, addressing, and frame-check sequence; it includes timers, frame counts, and priority stacks. It also defines the MAC protocol and provides finite-state machines and state tables. It identifies the services provided by the MAC sublayer to the LLC sublayer, and the services provided by the physical layer to this MAC

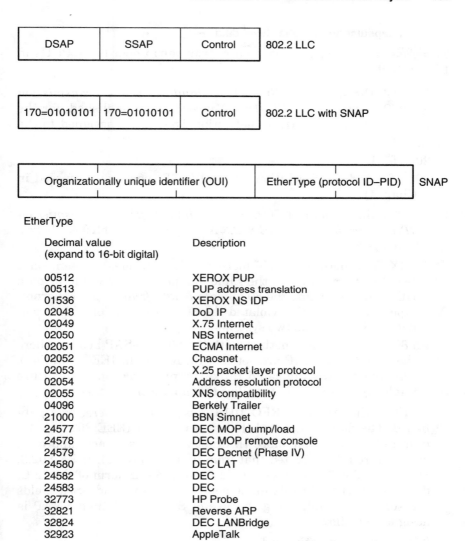

Figure 3.6 SNAP extension (examples). [*Note:* OUI is set to 0 for EtherTypes (refer to text).]

sublayer. These services are defined in terms of service primitives and associated parameters. It also defines the physical layer functions of symbol encoding and decoding, symbol timing and latency buffering, and the 1-Mb/s and 4-Mb/s twisted-pair attachments of the station to the medium. The specification also describes the service primitives between the MAC layer and the physical layer. This standard is representative of IBM's token-passing ring LAN. The 802.5 standard is for the 4 Mb/s rate and does not cover the 16 Mb/s rate supported by IBM's token ring.

3.2.6 Encapsulation of higher-level data

Three methods exist to encapsulate (transmit) upper-layer PDUs in LLC, as follows:

1. In 1984, the standard RFC 894, *A Standard for the Transmission of IP Datagrams over Ethernet Networks,* was published, for the use of Ethernet-type (DIX) networks. The values assigned to the *type-field* were:

   2048 for IP packets

   2054 for address resolution protocol (ARP) packets (discussed in Chap. 4)

2. In 1985, the standard RFC 948, *Two Methods for the Transmission of IP Datagrams over IEEE 802.3 Networks,* was published. It listed two options:

   A DIX Ethernet-compatible method, where frames are sent on a IEEE 802.3 network in the same manner as they would be on a DIX Ethernet, using the 802.3 *data-length field* as the Ethernet-type field. While this violated the 802.3 rules, it provided compatibility across the two systems.

   An 802.2 LLC Type 1 method, with the 802.2 L-SAP header where the SSAP and DSAP are set to 6 (01100000 in IEEE notation). This option was indicated to be the preferred one. All future implementations were to utilize this approach.

3. In 1987, the standard RFC 1010, *Assigned Numbers,* was published. The document noted that as a result of IEEE 802.2 evolution and the need for more Internet addresses, a new approach was required. It indicated that from that point on, all IEEE 802.3, 802.4, and 802.5 networks should use the SNAP form of the LLC, discussed earlier. In this implementation, DSAP and SSAP fields are set to 170 (indicating the use of SNAP) and then SNAP is assigned as follows:

   0 (zero) as organization code

   EtherType field (see Fig. 3.6)

   - 2048 for IP packets
   - 2054 for ARP packets
   - 32821 for reverse ARP packets (discussed in Chap. 4)

In 1988, this approach was formalized in RFC 1042, *A Standard for the Transmission of IP Datagrams over IEEE 802 Networks.* There are still some TCP/IP implementations that use the older L-SAP method (RFC 948), but these implementations do not allow communication with the more recent systems.

Preamble 56	SD 8	Destination address 48	Source address 48	PT 16	MAC data unit (≤1500)	FCS 32			Ethernet
Preamble 56	SD 8	Destination address 48	Source address 48	Length 16	MAC data unit (≤1500)	FCS 32			IEEE 802.3 MAC
Preamble (≥1 octet)	SD 8	FC 8	Destination address 48	Source address 48	MAC data unit (≤8174)	FCS 32	ED 8		IEEE 802.4 MAC
	SD 8	AC 8	Destination address 48	Source address 48	MAC data unit	FCS 32	ED 8	FS 8	IEEE 802.5 MAC

Figure 3.7 Comparison between MAC frames. SD = start delimiter (=10101011 in 802.3). ED = end delimiter. FC = frame control. AC = access control. FS = frame status. PT = protocol type. Field lengths shown in bits, except for information field, shown in octets.

3.2.7 Comparison of MAC layers

Figure 3.7 compares the format of the three MAC layers described above.

References

1. B. Bhushan, "Twisted Pair Wiring Comes of Age for LANs," *Networks In-Depth,* July 1991.
2. D. Zwicker et al., "Ethernet over Twisted Pair: Not as Easy as It Looks?" *Telecommunications,* June 1990, pp. 23 ff.
3. IBM, *TCP/IP Tutorial and Technical Overview,* June 1990, Document GG24-3376-01.

Figure 3.7 ...

References

LAN Basics: First-Generation Upper Layers

Upper-layer protocols are needed for both station-to-station communication within a LAN and interconnection of LANs. This chapter focuses on a de facto set of upper-layer protocols, namely TCP/IP. Some OSI views are also presented.

4.1 Connectionless versus Connection-oriented Communication

At any layer of the OSIBRM discussed in Chap. 2 except the physical layer, two basic forms of operation (service) are possible: Connection-oriented mode and connectionless mode.

A *connection-oriented service* involves a connection establishment phase, a data transfer phase, and a connection termination phase. This implies that a logical connection is set up between end systems prior to exchanging data. These phases define the sequence of events ensuring successful data transmission. Sequencing of data, flow control, and transparent error handling are some of the capabilities inherent with this service mode (see Fig. 4.1). One disadvantage of this approach is the delay experienced in setting up the connection. Traditional carrier services, including circuit switching, X.25 packet switching, and early frame relay service, are examples of connection-oriented transmission; LLC 2, discussed earlier, is also a connection-oriented protocol.

In a *connectionless service,* each PDU is independently routed to the destination. No connection-establishment tasks are required, since each data unit is independent of the previous or subsequent one. Hence, a connectionless service provides for transfer of data units

Connection-mode transmission

Connection establishment

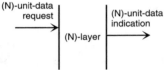

Figure 4.1   Connection-oriented and connectionless operation.

(cells, frames, or packets) without regard to the establishment or maintenance of connections (see Fig. 4.1). The basic MAC/LLC (i.e., LLC 1) transfer mechanism of a LAN is connectionless; so is a service such as SMDS. Clearly, each unit of data must contain the addressing information and the data itself. In the connectionless-mode transmission, delivery is uncertain because of the possibility of errors. Connectionless communication shifts the responsibility for the integrity to a higher layer, where the integrity check is done only once, instead of being done at (every) lower layer.

While the original Open Systems Interconnection Basic Reference Model described in ISO 7498 was connection-oriented, ISO subsequently extended it to provide connectionless service by issuing an addendum to that standard. Table 4.1 provides a comparison between the two modes.

4.2   Internetworking History and Goals

As indicated above, upper-layer protocols become critical when dealing with internetworking.

4.2.1   Background

Internetworking hides the details of network hardware from the user and permits computers to communicate independent of their physical

TABLE 4.1   Comparison of Connectionless and Connection-oriented Communication

	Connection-oriented	Connectionless
Packet sequencing	Yes	No
Flow control	Yes	No
Acknowledgments	Yes	No
Protocol	Complex	Simple
Packet handling	Packet layer sets up logical channel	Packets are sent independently
	Each packet has a logical channel identifier	Each packet has complete addressing information
	Same virtual circuit for duration of the call	Packets can take totally different routes
Typical network layer protocol	ISO 8202 [(X.25 packet level protocol (PLP)]	Internet IP; ISO 8473 connectionless network protocol (CLNP)

network connections. Compatible protocols are required to interconnect distinct networks. Such protocols aim at making these networks
appear as a single cohesive system to the end user. Interconnected
networks are also called an *internetwork* or an *internet*. For the purpose of this book we dispense with the additional jargon and refer to
these networks as *interconnected networks*.

There exist widely deployed pre-OSI protocol suites used to facilitate network interconnection. A de facto standard is TCP/IP. TCP is a
transport layer protocol, and IP is a network layer protocol. TCP
takes care of the integrity, and IP moves the data. TCP is connection-
oriented, while IP is connectionless. TCP/IP principles can be applied
to both the general interconnection problem and to the LAN environment. *Of course, other vendor-specific protocol suites are available.*
Also, there is some commercial movement toward OSI-based LAN
stacks; these are discussed briefly toward the end of this chapter; for
additional information, see Minoli.[1]

U.S. government agencies were the first to realize the importance
of internetworking technology. The U.S. government funded research
over the past quarter century to make possible national internetworked communications systems. In the early 1970s the U.S. Defense
Advanced Research Projects Agency (DARPA) funded work which
developed network standards specifying the details of how computers
communicate. These protocols are commonly referred to as the
Internet protocol suite. Internet is the proper name of an actual network, while "internet" refers to a generic internetworked environment. Internet is a collection of several thousand packet-switched
networks, located principally in the United States; but it also includes
systems in other parts of the world. Interconnection is via TCP/IP.

The TCP/IP architecture and protocols as we know them today, acquired their form in the late 1970s. In the late 1960s, DARPA initiated an effort to develop the first packet-switching network, known as ARPANET. The successful implementation of networking technologies raised the possibility of interconnecting the ARPANET with other types of packet networks. The solution to this challenge was developed as part of research programs sponsored by DARPA, and resulted in a collection of protocols based on TCP and IP (for ARPANET, the transition to TCP/IP technology was completed in 1983). The decision on the part of DARPA to make an implementation available at low cost encouraged the academic community to install the technology. These protocols, along with others developed during the years, are known as the TCP/IP or Internet protocol suite and are now very common in the LAN environment.

The Internet Activities Board (IAB) provides the focus for research and development in support of the TCP/IP suite. In the early stages of the Internet research program, only a few researchers worked to develop network interconnection protocols. Over time, the size of this activity increased until, in 1979, it was necessary to form an informal committee to guide the technical evolution of the TCP/IP suite. This group was called the Internet Configuration Control Board (ICCB). In 1983, the Defense Communications Agency, then responsible for the operation of the ARPANET, declared the TCP/IP protocol suite to be the standard for ARPANET and for the Defense Data Network (DDN). All systems on the network converted from the earlier Network Control Program to TCP/IP. Late that year, the ICCB was reorganized around a series of task forces considering different technical aspects of internetworking. The reorganized group was named the IAB. The IAB is now the coordinating committee for Internet design, engineering, and management. The IAB is an independent committee of researchers and professionals with a technical interest in the evolution of the Internet system in general, and TCP/IP in particular. Membership changes with time to adjust to the needs of the Internet system and the concerns of the U.S. government, universities, and industrial sponsors.

All decisions of the IAB are made public. The principal vehicle by which the decisions are propagated is the *request for comment* (RFC) series. These documents describe experimental protocols. A standard RFC starts out as a *proposed standard* and may be promoted as a *draft standard,* and finally *standard,* after suitable review, comment, implementation, and testing. Each RFC has two attributes: a *state,* which indicates the document's level of standardization, and a *status,* which indicates the level of support the Internet community must accord the document. States are Internet Standard, Draft Standard, Proposed Standard, Experimental Protocol, and Historical Protocol.

Statuses are Required (all hosts in the system must implement it), Recommended (all hosts in the system are encouraged to implement it), Elective (hosts in the system may decide whether to implement it or not), and Not Recommended.

4.2.2 Layering

Communication functions can be mapped to the OSI layering model as described in Chap. 1; however, some prefer a four-layer partitioning of the communication functions, as seen in Fig. 4.2 and described below.

Application stratum. This refers to a user process either in support of a business function (such as inventory, payroll, reservations, etc.) or of a communication function (such as e-mail transfer, file transfer, network directory, etc.). (This includes OSI layers 7, 6, and 5.)

Transport stratum. This refers to the ability to assure reliable end-to-end data transfer between two networks, meeting a specified level of integrity and quality of service. Integrity is particularly important when end-to-end communication has to rely on multiple intermediate networks. TCP supports this layer. (This corresponds with OSI layer 4.)

Internetworking stratum. This refers to the ability to send data from one network to another network. It deals with network and device addressing issues (since devices may be identified using different

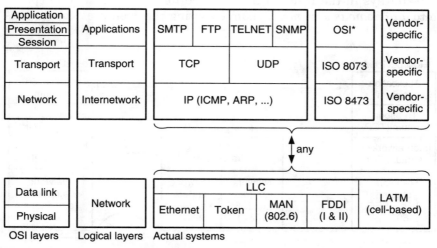

Figure 4.2 A variety of LAN architectures compared with the OSI model and a logical view of required functions. *This includes FTAM, MHS, etc., at the application layer; ISO 8823 at the presentation layer; and ISO 8327 at the session layer. (*Note:* See Chap. 2 for a more inclusive version of this protocol map.)

schemes in the two networks), multiplexing, and connection selection and routing. IP supports this layer. Networks are connected to one another by a *gateway* (also known in this context as an *IP router*). (This corresponds with OSI layer 3.)

Networking stratum. This refers to the ability to form a basic self-contained network and to be able to send data from one user on such a network to another user on the same network (In an obvious deviation from OSI nomenclature, this corresponds to OSI layers 2 and 1. Such a variation from the OSI nomenclature is not capricious, because, for example, a LAN (which, at the core, supports OSI layers 1 and 2) is called a *network* (LA*N*). BISDN (which supports the equivalent of layers 1 and 2) is also called a network (BISD*N*). Hence, systems supporting layers 1 and 2 can be considered a network.)

4.2.3 Gateways

As discussed in Chap. 1, LAN interconnection can be achieved with bridges, routers, and application layer gateways. In that context, gateways are used when interconnecting two completely different networks such as an IBM PC network and DECnet. In TCP/IP environments, the term *gateway* refers to a hardware device used to connect two physical networks, of the same or different type, running the IP protocol (see Fig. 4.3). This device is also called an *IP router*. The gateway must, among other things, be able to fragment IP packets to match the maximum size supported by an outbound link.

Gateways in the Internet are grouped for administrative purposes into *autonomous systems*. Dynamic routing relies on protocols to con-

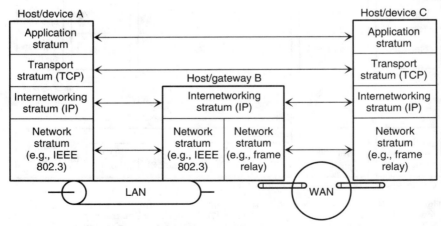

Figure 4.3 IP routing from a protocol stack perspective.

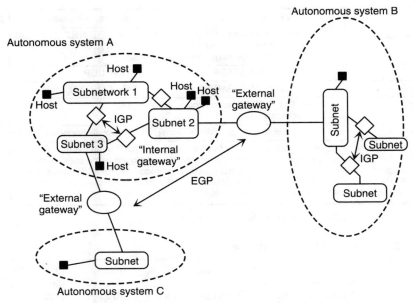

Figure 4.4 Internal and external gateways in Internet. IGP = internal gateway protocol; EGP = external gateway protocol.

vey the routing information around the interconnected networks. When a topology change occurs in the (interconnected) network, the routers directly involved in the change are responsible for propagating this information to the other routers in the system. Gateways within an autonomous system communicate with each other using one of a number of dynamic routing protocols, known collectively as *interior gateway protocols*. This communication is required to dynamically update the routing information in each gateway, to reflect real-time conditions of the topology. Exchange of routing information between gateways of different autonomous systems requires *exterior gateway protocols* (see Fig. 4.4).

Hence, a TCP/IP gateway must implement IP, the Internet Control Message Protocol (ICMP), one or more internal gateway protocols, and, optionally, if the gateway connects multiple autonomous systems, the exterior gateway protocol.

4.3 TCP/IP Protocol Suite

The basic TCP/IP protocol suite is shown in Fig. 4.5 for both LAN and WAN applications. There are about 100 protocols in the Internet suite. Table 4.2 provides additional information about some key Internet protocols.

	LAN environment	WAN environment
Layer 7–Layer 5	Application-specific protocols such as TELNET (terminal sessions), FTP and SFTP (file transfer), SMTP (e-mail), SNMP (management), and DNS (directory)	
Layer 4	TCP, UDP, EGP/IGP	TCP, UDP
Layer 3	IP, ICMP, ARP, RARP	IP, ICMP, X.25 PLP
Layer 2	LLC; CSMA/CD, token ring, token bus	LAP-B
Layer 1	IEEE 802.3, .4, .5 (PMD portions)	Physical channels

SFTP = simple file transfer protocol; FTP = file transfer protocol; SMTP = simple mail transfer protocol; SNMP = simple network management protocol; DNS = domain name service; UDP = user datagram protocol; ICMP = internet control message protocol; ARP = address resolution protocol; RARP = reverse address resolution protocol; EGP = external gateway protocol; IGP = internal gateway protocol; PMD = physical medium dependent.

Figure 4.5 TCP/IP-based communication: key protocols.

TABLE 4.2   Key Internet Protocols (Partial List)

—	IAB Official Protocol Standards	Req	1280
—	Assigned Numbers	Req	1060
—	Host Requirements—Communications	Req	1122
—	Host Requirements—Applications	Req	1123
—	Gateway Requirements	Req	1009
IP	Internet Protocol	Req	791
	IP amended by:		
—	IP Subnet Extension	Req	950
—	IP Broadcast Datagrams	Req	919
—	IP Broadcast Datagrams with Subnets	Req	922
ICMP	Internet Control Message Protocol	Req	792
IGMP	Internet Group Multicast Protocol	Rec	1112
UDP	User Datagram Protocol	Rec	768
TCP	Transmission Control Protocol	Rec	793
TELNET	Telnet Protocol	Rec	854, 855
FTP	File Transfer Protocol	Rec	959
SMTP	Simple Mail Transfer Protocol	Rec	821
MAIL	Format of Electronic Mail Messages	Rec	822
CONTENT	Content Type Header Field	Rec	1049
NTP	Network Time Protocol	Rec	1119
DOMAIN	Domain Name System	Rec	1034, 1035
DNS-MX	Mail Routing and the Domain System	Rec	974
SNMP	Simple Network Management Protocol	Rec	1157
SMI	Structure of Management Information	Rec	1155
MIB-II	Management Information Base-II	Rec	1213
EGP	Exterior Gateway Protocol	Rec	904
NETBIOS	NetBIOS Service Protocol	Ele	1001, 1002
ECHO	Echo Protocol	Rec	862
DISCARD	Discard Protocol	Ele	863
CHARGEN	Character Generator Protocol	Ele	864
QUOTE	Quote of the Day Protocol	Ele	865
USERS	Active Users Protocol	Ele	866
DAYTIME	Daytime Protocol	Ele	867
TIME	Time Server Protocol	Ele	868

Req = required; Rec = recommended; Ele = elective.

A TCP/IP LAN application involves (1) a user connection over a standard LAN environment (IEEE 802.3, .4, .5 over LLC); (2) software in the PC or server implementing the IP, TCP, and related protocols, and (3) programs running in the PCs or servers to provide the needed application. (The application may use other higher-layer protocols for file transfer, network management, and so on.) The most widespread traditional TCP/IP user applications are electronic mail, file transfer, and access to hosts on remote networks.

4.4 IP Protocol

In a TCP/IP environment, IP provides the underlying mechanism to move data from one end system on one LAN to another end system on the same or different LAN. IP makes the underlying network transparent to the upper layers, TCP in particular. It is a connectionless packet delivery protocol, where each IP packet is treated independently. In this context, packets are also called *datagrams.* IP provides two basic services: addressing and fragmentation and reassembly of long packets. IP adds no guarantees of delivery, reliability, flow control, or error recovery to the underlying network other than the data link layer mechanism already provides. IP expects the higher layers to handle such functions. IP may lose packets, deliver them out of order, or duplicate them; IP defers these contingencies to the higher layers (TCP, in particular). Another way of saying this is that IP delivers on a "best-effort basis." There are no connections, physical or virtual, maintained by IP. To provide its services, IP employs four key header fields:

- *Type of service.*   Parameters set by the end station specifying, for example, expected delay characteristics, expected reliability of path, etc.

- *Time to live.*   Parameter used to determine the packet's lifetime in the interconnected system

- *Options.*   Parameters to specify security, timestamps, and special routing

- *Header checksum.*   A two-octet field used by IP to determine packet integrity

A more detailed discussion of IP formats is provided in later subsections.

4.4.1 IP addresses

IP addressing provides a basic way to identify a device on an interconnected network. IP requires that the network manager properly set up an IP address for each device on the network, so that it can receive information from a remote sender. An IP address comprises a *network address* used to identify the network to which a device (also called host) such as a PC terminal or computer is connected and an *identifier for the device itself*. It is 32 bits in length. An IP address can be represented as:

AdrType I netID I hostID

A packet (protocol data unit) coming down the stack of a PC or host connected to the network contains the IP address of the origination as well as the address of the destination. If the destination device is on the same network as the originating device, the packet will be directly "absorbed" by the destination. If the network is different from the network of the originating device, the packet must first be routed to that remote network (or some intermediate network), where it will be "absorbed" by the intended device.

Each IP address must be unique. This is because of DARPA's requirement to be able to interconnect a multitude of networks with a worldwide backbone (the Internet). To supervise the uniqueness of IP addresses, a central body in each country that implements IP-based networks is responsible for administering and distributing IP addresses. Overall responsibility for IP addresses rests with the Internet Assigned Number Authority (IANA) group.

There are five standardized ways describing how the 32 address bits are to be allocated; these are referred to as classes A, B, C, D, and E (see Fig. 4.6). These classes are used to accommodate different requirements in terms of enterprise size.

- *Address class A.* Class A uses the first bit of the 32-bit space (bit zero) to identify it as a class A address; this bit is set to 0. Bits 1 to 7 represent the network ID, and bits 8 to 31 identify the PC, terminal, or host on the network. Clearly every device and every network has a unique identifier. This address supports $2^7 - 2 = 126$ networks and approximately 16 million (2^{24}) devices on each network. IP prohibits the use of an all 1s or all 0s address for both the network and the device ID (which is the reason for subtracting 2).

- *Address class B.* Class B uses the first two bits (bit zero and one) to identify it as a class B address. These bits are set to 10. Bits 2 to 15 are used for network IDs, and bits 16 to 31 are used for device IDs. This address supports $2^{14} - 2 = 16,382$ networks and $2^{16} - 2 = 65,134$ devices on each network.

- *Address class C.* Class C uses the first three bits to identify it as a class C address. These bits are set to 110. Bits 3 to 23 are used for network IDs, and bits 24 to 31 are used for device IDs. This

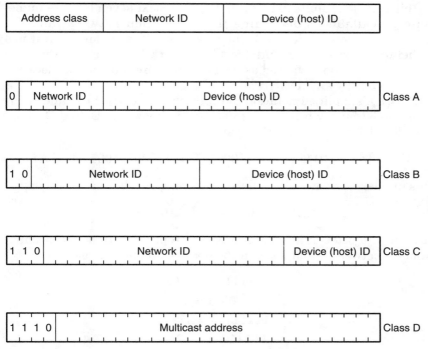

Figure 4.6 Structure of IP address.

address supports over 2 million (2^{21}) networks and $2^8 - 2 = 254$ devices on each network.

- *Address class D.*   This class is used for broadcasting (multiple devices receive the same IP-level packet).
- *Address class E.*   Reserved for future use.

An address with all bits equal to zero represents "this" (this network or local host); an address with all bits equal to one stands for "all" (all networks or hosts). Large organizations (universities, *Fortune* 100 companies, etc.) are typically granted class B addresses. Organizations with a large number of networks are assigned class C. Early participants in the Internet have class A addresses. Class A addresses are assigned to networks with a large number of devices; class C addresses are assigned to networks with a small number of devices. An extension to the address mechanism will be required by 1994 when the address space administered by IANA is expected to be exhausted.

The address class encoding described above facilitates efficient movement of data over the network, from an address-decision point of view. As soon as the address bits are boundary-synchronized, the first bit (bit zero) is examined. If it is 0, then the next seven bits are immediately available to determine the destination network (and the route). If the zeroth bit was 1, then the first bit is examined. If it is a 0, then the next fourteen bits are the network ID. Note that the class identification code is viewed as part of the address itself, but does not actually identify any network.

IP uses a simplified notation to represent, for ease of use and reference, the 32 binary bits.[2,3] This notation is known as Dotted Decimal Notation (DDN). Consider the example

IP Address = 01111110011000011111111000111001 (class A).

This number is broken down into four octets, as follows:

01111110-01100001-11111110-00111001.

In turn, the octet is assumed to be the representation of a decimal number (between 0 and 255). In the above example, one has

$$01111110 = 126$$
$$01100001 = 97$$
$$11111110 = 254$$
$$00111001 = 57$$

Finally, the IP address is represented as 126.97.254.57. In fact, this notation can be used directly when LAN manager specifies routing tables, since there is an internal translation function, so that the DDN number is automatically translated to binary. For example, the thirteenth device is represented as 13; the thirtieth device as 30; etc. An example of actual IP addresses is shown in Table 4.3. Figure 4.7 depicts a corporate network (partial) connected with Internet. The

TABLE 4.3   Illustrative IP Addresses for Bell Communications Research (Partial List)

	Class B addresses
bellcore.com	128.96.xyz.abc
bellcore-cisco.bellcore.com	128.96.34.1
wbdsn5e.bellcore.com	128.96.34.2
...	
thumper.bellcore.com	128.96.41.1
...	
orion.bellcore.com	128.96.43.1
...	
...	

TABLE 4.3   **Illustrative IP Addresses for Bell Communications Research (Partial List)** *(Continued)*

Class B addresses (*Continued*)	
pyuxp. cc.bellcore.com	128.96.96.1*
dasher. cc.bellcore.com	128.96.96.2*
...	
bcr.cc.bellcore.com	128.96.96.105*
netcom.cc.bellcore.com	128.96.96.106*
...	
nvuxr.cc.bellcore.com	128.96.96.118*
...	
pyahub-cisco. cc.bellcore.com	128.96.96.254*
pyumv.bellcore.com	128.96.97.1
pymvsgp.bellcore.com	128.96.97.2
pymvstst.bellcore.com	128.96.97.3
...	
deal.osn.bellcore.com	128.96.98.40†
repon.osn.bellcore.com	128.96.98.41†
...	
georgemac.osn.bellcore.com	128.96.98.123†
nv750.bellcore.com	128.96.100.1
caesar.bellcore.com	128.96.100.2
dgpnwkst.bellcore.com	128.96.100.3
...	
pyahub-cisco-gw.cc.bellcore.com	128.96.101.2‡
tetd.cc.bellcore.com	128.96.101.3‡
pya2-backbone-ods-1.cc.bellcore.com	128.96.101.4‡
...	
groucho.bellcore.com	128.96.102.1
...	
...	
bulldog.bellcore.com	128.96.115.55
...	
...	
nma-eg.bellcore.com	128.96.168.61
Class C addresses	
pronet-tiu.bellcore.com	192.4.4.4
...	
pronet-monitor.bellcore.com	192.4.4.253
...	
tuesday.bellcore.com	192.4.13.1
...	
wodehouse.bellcore.com	192.4.13.254
...	

Subnetwork mask for class B address: 255.255.255.0 (i.e., third octet designates subnetwork IDs)

*.cc : mnemonic name subnetwork.
†.osn : mnemonic name of another subnetwork.
‡Mnemonic name of subnetwork is not unique (although the actual address is unique).

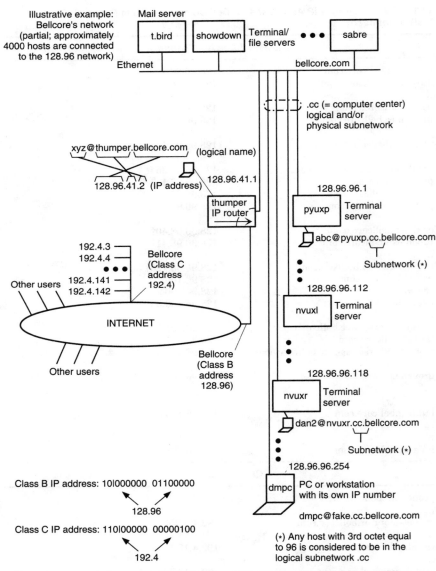

Figure 4.7 An example of a corporate network connected to the Internet.

".com" suffix refers to a *commercial* institution; an ".edu" suffix would refer to an *educational* institution. Figure 4.8 depicts an example of mail receipt over such network: a corporate user sent mail to a nationwide work group connected over the Internet; this author (dan2), also on the corporate network, receives a copy of the mail.

(line 1) From showdown.bellcore.com!tkh Fri May 1 15:32:22 1992

(line 2) Received: from thumper.bellcore.com by tbird.cc.bellcore.com with SMTP id AA2983 5.65c/IDA-1.4.4 for <dan2@nvuxr.cc.bellcore.com>); Fri, 1 May 1992 15:32:22 0

(line 3) Received: from sabre.bellcore.com by thumper.bellcore.com (4.1/4.7) id <AA10496> for dan2@nvuxr.cc.bellcore.com; Fri, 1 May 92 15:31:51 EDT

(line 4) Received: by sabre.bellcore.com (5.57/Ultrix2.4-C) id AA14990; Fri, 1 May 92 15:25:35 EDT

(line 5) Return-Path: <tkh@sabre.bellcore.com>

(line 6) Received: by showdown (4.1/4.7) id AA01085; Fri, 1 May 92 15:32:04 EDT

(line 7)Date: Fri, 1 May 92 15:32:04 EDT

(line 8) From: tkh @sabre.bellcore.com (Thomas Hxyzpqr)

(line 9) X-Station-Sent-From: showdown.bellcore. com

(line 10) Message-Id: <9205011932.AA01085@showdown>

(line 11) To: latm-sig@thumper.bellcore.com

(line 12) Subject: Signaling meeting

(line 13)....

(line nnn) See all of you soon...

Commentary:

From: Mail sender was on showdown.bellcore.com

- Full address = tkh@showdown.bellcore.com (line 1)
- showdown - tkh's machine
- bellcore.com = Bellcore's backbone network (alias)
- 128.96 = Class B IP address
- Preferred return address for sender is tkh@sabre.bellcore.com (line 8)

To: Mailing list is latm-sig@thumper.bellcore.com (line 11)

List cannot be resolved on showdown, so it must be resolved elsewhere; sabre is local mail server

- You read the path from the bottom up
- showdown sends mail to sabre (line 4)
- sabre sends mail to thumper (line 3)
- thumper expands list (line 11)
- thumper sends mail (i.e., mail with real addresses affixed) to tbird which is Bellcore's mail gateway (line 2)
- tbird will send the mail for remote users over the Internet (based on the expanded mailing list)
- dan2 (Dan Minoli) is at nvuxr.cc.bellcore.com (nvuxr = machine; cc = subnetwork); gets copy of mail via tbird

Figure 4.8 Outbound e-mail session over the Internet.

Figure 4.9 shows the receipt of e-mail sent from a remote user over the Internet.

One drawback of this addressing scheme is that if a device or host moves from one network to another, its IP address must be changed. Non-IP based networks (AppleTalk, DECnet, NetWare, etc.) do not employ this addressing scheme.

Plans are beginning to be made to extend the 4-octet IP address. The current address only supports about 2 million addresses. The IAB expects this address space to become exhausted by 1995. A 1992 proposal to deal with this problem by replacing IP with ISO 8473

(line 1) From srv.pacbell.com!yalin Fri May 1 16:43:56 1992

(line 2) Received: from thumper.bellcore.com by tbird.cc.bellcore.com with SMTP id AA0675 (5.65c)/IDA-1.4.4 for <dan2@nvuxr.cc.bellcore.com>); Fri, 1 May 1992 16:43:56 0

(line 3) Received: from ns.PacBell.COM by thumper.bellcore.com (4.1/4.7) id <AA16847> for dan2@nvuxr.cc.bellcore.com; Fri, 1 May 92 16:43:23 EDT

(line 4) Received: from srv.PacBell.COM (mother.srv.PacBell.COM) by ns.PacBell.COM (4.1/) id AA21334; Fri, 1 May 92 13:32:02 PDT

(line 5) Received: from pop.srv.PacBell.COM by srv.PacBell.COM (4.1/SMI-4.0) id AA06384; Fri, 1 May 92 13:32:00 PDT

(line 6) Received: from firsco by pop.srv.PacBell.COM (4.1/SMI-4.1) id AA 15688; Fri, 1 May 92 13:31:59 PDT

(line 7) Date: Fri, May 92 13:31:59 PDT

(line 8) From: yalin@srv.pacbell.com (Axy Lxy)

(line 9) Message-id: <9205012031.AA15688@pop.srv.PacBell.COM>

(line 10) To: tkh@sabre.bellcore.com

(line 11) Subject: Re: signaling meeting

(line 12) Cc: latm-sig@thumper.bellcore.com

Commentary:
- From: Network server at remote firm: ns.Pacbell.COM (line 4)
- Mail received over the Internet by thumper.bellcore.com (= secure Bellcore node) (line 2)
- Mail sent to t.bird.cc.bellcore.com (line 3)
- thumper expends mailing list and forwards message (cc copy) to dan2@nvuxr.cc.bellcore.com (line 12 and line 2)

Figure 4.9   Inbound e-mail session over the Internet.

(CLNP), which allocates 20 octets for addressing, was met with major resistance by the user community and was soon withdrawn. Some expect the IP address to be gracefully extended to 8 octets. A variety of methods are under study.

4.4.2   Subnetwork addressing

Usually organizations have *subnetworks* which comprise the larger network (independent of the address class they may have). This is done for performance or administrative reasons. IP addresses can become inflexible when it comes to making changes to local network configurations. Those changes might occur when (1) a new type of physical network needs to be installed at a location; (2) there is growth in the number of devices (user terminals and hosts), requiring the segmentation of the local network into two or more networks; (3) there is growth in the site-to-site distances (for example, a group of people is moving into a building a few miles away), requiring the segmentation of the network into smaller networks, with gateways between them. One goal is to be able to perform these network redesign functions without having to be assigned new IP addresses.

The establishment of subnetworks can be done locally, while the whole network still appears to be one IP network to the outside world. As discussed above, the IP addresses consist of the pair

<network address><host address>

IP allows a portion of the host-device field to be used to specify a subnetwork (the network ID portion cannot be changed). Subnetworks are an extension to this scheme by considering a part of the <host address> to be a "local network address," that is, a subnetwork address. IP addresses are then interpreted as

<network address><subnetwork address><host address>

For example, in class A addressing, a subset of the bits from bit 8 to bit 31 could be employed for subnetwork identification. The partition of the original <host address> into a <subnetwork address> and <host address> part can be done by the local administrator without restriction. However, once this partition has been established, it must be used consistently throughout the whole local network. Also, whereas bits can in theory be used freely, it is best to employ a contiguous set to represent the subnetwork.

Submasks are used to describe subnetworks; they tell devices residing on the network how to interpret the device ID portion of the IP address. The address checking software in each device is informed via the submask not to treat the device ID exclusively as a device identifier but as a subnetwork identifier followed by a (smaller) device identifier. Naturally, since the address space is finite, there is a tradeoff between the number of subnetworks that can be supported and the number of devices on each subnetwork.

The mask contains a bit for each bit in the IP address, although the "active ingredients" portion of the mask is really only contained in the section describing the device ID. If the bit is set in the IP address mask, the corresponding bit is to be treated as a subnetwork address. All unaffected bits in the left portion of the mask are set to 1. Bits set to 0 represent the actual extent of the device address. For example, consider the class B IP address

10 | 11111001100001 | 1111111000111001

(the symbol "|" is only used for visual convenience; it is not part of the address). Normally the last 16 bits are used to identify devices on the (single) network (whose identity is 11111001100001). A mask of the form

11 | 11111111111111 | *111111111*000000

implies that the first 10 binary positions of the device ID field are used to represent subnetworks. In addition to using the DDN as an address-representation scheme, it is also used to describe the submasks. Here, though, the representation is only a shorthand for the bit pattern of the mask, and clearly not an address. A LAN manager would specify the mask just described as

11111111-11111111-11111111-11000000

or, compactly,

255.255.255.192

Class A addresses allow only the following masks: 255.xyz.abc.ghi (as seen below, not all 255 combinations actually make sense). Specifically, bits 0 to 7 cannot be altered to represent subnetworks and so are shown in the mask as a string of eight ones. This arrangement allows considerable flexibility in terms of the number of subnetworks that can be defined. Class B addresses allow only the following submasks: 255.255.xyz.abc. This class affords a fair degree of flexibility in terms of subnetworks. Class C addresses allow only the mask 255.255.255.xyz. It offers limited flexibility in terms of subnetworks; however, it still allows a balance between subnetworks and the number of devices per subnetwork.

With class B, a typical mask is 255.255.255.0. This corresponds to 11111111-11111111-11111111-00000000, meaning that the first two octets represent the network ID, the next octet represents a subnetwork ID, and the last octet represents a device ID. Up to 254 subnetworks, each with up to 254 devices, are supported. This octet-based partition for the network ID, the subnetwork ID, and the device ID, is easy to parse, making routing decisions simple. An example of a network address could be 128.79.xyz.abc (again, note that the IP network address is assigned by the IANA). If one wanted only the last six bits to represent device addresses, then the encoding would be 255.255.252.192 (i.e., the last octet would be 11000000).

As hinted earlier, not all masks are easy or useful to deal with. In the example just given, a mask of 255.255.255.17 would imply the following binary pattern:

11111111-11111111-11111111-00010001.

This would imply that bits 27 and 31 represent the subnetwork ID. Such an address would be difficult to parse. It is preferable to use representations which result in a contiguous stream of 0s and 1s.

All the devices in a subnetwork must use and implement the same address mask. However, an IP router node uses the full four-octet address to route to the destination network, and so it is not required to interpret the subnetwork field on the transmit side.[2,3]

4.4.3   Use of the IP address

As SDUs make their way down a TCP/IP LAN stack, they are enveloped with headers and trailers at each layer (thereby becoming a PDU). An IP address is included in the IP PDU. This, in turn, is given to the LLC, which, as seen earlier, adds its own destination SAP address, enveloping the IP address. In turn this frame is given to the MAC layer, which adds its own 48-bit address, enveloping the LLC and IP address. As data flows over the medium to another device (whether an actual user, or a bridge, router, gateway, etc.), the readily accessible MAC-level address is used to determine if the data is intended for the physical device in question. Once such a determination is made, the LLC-level address is used to determine if the data reaching the device in question is meant for one or another logical link which may have been established by (software) resources in the device. If this device is another user station on the LAN, the IP address is not functionally needed, although a TCP/IP application such as SMTP may require a value in the field anyway. (If the network is strictly private, a dummy value can be used.) However, if the device receiving the information is a router to support interconnection with a remote network, the IP address is instrumental in enabling the router to identify which network is required, and, after such determination, to select the physical port or line over which the data needs to be transferred.

4.4.4   IP routing tables

A key function of the internetworking stratum (the IP layer) is to support routing, properly called *IP routing* in this context. Any host or machine implementing IP can be a gateway. An IP router (gateway) can be a normal host running TCP/IP, as the gateway functionality is included in the basic IP protocol.[4] In other words, gateways used for interconnecting different physical networks employ routing as a basic mechanism, meaning that the IP routing capability is sufficient to perform the gateway function. Figure 4.3 depicts this functionality from a protocol stack point of view.

Basic IP gateways, also known as *gateways with partial routing information,* only have information about the devices directly attached to the physical networks to which this gateway is attached. The IP routing table contains information about the locally attached networks and IP addresses of other gateways located on these networks, in addition to the networks they attach to. The table can be extended with information on IP networks that are further away, and can contain a default route, but it still remains a table with limited information. Hence, this kind of gateway is called a gateway with partial routing information.

Gateways with partial information are characterized by the following:

- They do not have knowledge of all interconnected networks.

- They allow local autonomy in establishing and modifying routes.

- Routing entry errors in one gateway may introduce inconsistencies, thereby making part of the network unreachable.

Some configuration require more than just the basic routing function; these configurations require a gateway-to-gateway communication mechanism to relay routing information, as discussed earlier. A more sophisticated gateway system is required if:

- The gateway needs to know routes to all possible IP networks.

- The gateway needs to have dynamic routing tables. Dynamic routing tables are kept up to date with minimal or no manual intervention.

- The gateway has to be able to convey local changes to other gateways.

These more advanced forms of gateways use protocols to communicate with each other. These are known as *full-function gateways.*

From a maintenance perspective there are two types of gateway: *core gateways* and *noncore gateways.* Core gateways are maintained by the Internet Network Operations Center and provide reliable routes to all Internet subnetworks. They use protocols such as RFC 869, RFC 823 (gateway-to-gateway protocol), and RFC 904. Noncore gateways are created and administered by individual users. They use protocols such as RFC 1058 and RFC 904 (external gateway protocol).

If the destination device is attached to a network to which the source host is also attached, information can be sent directly by encapsulating the IP packet in the physical network frame. This is called *direct delivery-direct routing.* When the destination device is not on a network directly accessible to the source host indirect routing occurs. Here, the indented destination must be reached via one or more IP gateways. The address of the first of these gateways (the first hop) is called an *indirect route.* The address of the first gateway is needed by the source device in order to initiate the delivery of the information.

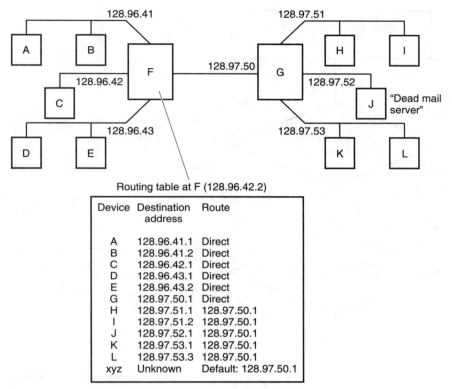

Figure 4.10 Example of routing table.

This simple view of the world makes the routing table relatively straightforward. A gateway keeps tracks of two sets of addresses in the *IP routing table* (see Fig. 4.10 for an example):

1. Devices attached to networks which are directly accessible. These devices have the same IP network ID address as the IP network ID of the source gateway-host itself.

2. For "indirect" hosts, the only knowledge required is the IP address of the "next gateway," that is, a gateway leading to the destination "IP network."

Additionally, the table contains a default route, which contains the (direct or indirect) route to be used in case the destination IP network is not otherwise identified. Figure 4.11 depicts the actual routing algorithm.

IP relaying is based on the network ID portion of the destination IP address. The device ID portion of the address plays no part at this stage. On the incoming side, arriving IP packets are checked to deter-

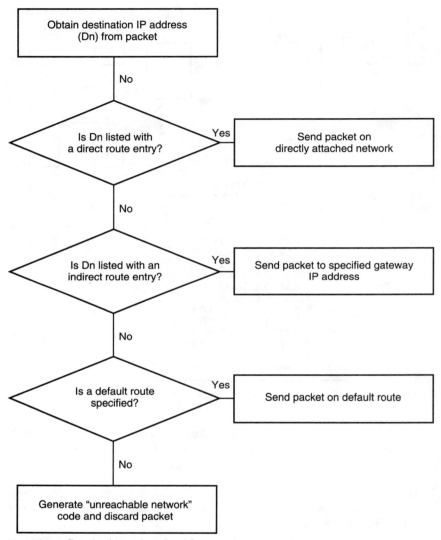

Figure 4.11 Gateway's routing algorithm.

mine if the IP address on the packet coincides with the IP address of the local network (that address can be thought of as being assigned to the IP router, rather than something more abstract as a network). If the addresses agree, the packet is passed up to the upper portions of the protocol stack. If the address does not agree, the IP router checks its routing tables to determine on which physical outgoing path the packet should be directed to (see Fig. 4.12).

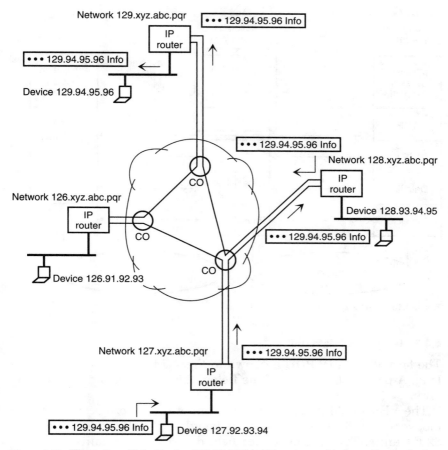

Figure 4.12 IP routing. *Note*: Device 127.92.93.94 wishes to send information to device 129.94.95.96.

The fundamental operation for gateways is as follows: An incoming IP packet that contains a "destination IP address," other than the local host or gateway IP address (or addresses), is treated as a normal outgoing IP packet. Any outgoing IP packet is subject to the IP routing algorithm of the gateway-host in question. The gateway-host selects the next hop for the packet (the next device-gateway-host to send it to) by checking its routing table. This new destination can be attached to any of the physical networks to which the gateway-host is connected. If this network is a different physical network from the one on which the gateway-host originally received the IP packet, then the net result is that the local gateway-host has forwarded the IP packet from one physical network to another (see Fig. 4.13).

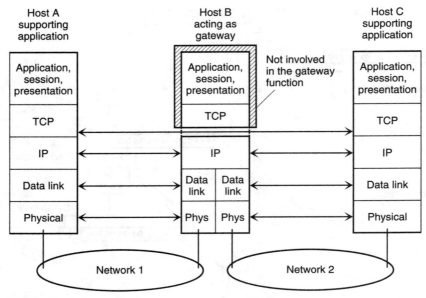

Figure 4.13 Gateway function.

4.4.5 IP protocol data unit

The format of an IP PDU is shown in Fig. 4.14. It is 20 or more octets long. A partial discussion of the fields, their purpose, and format follows.

The VERS field describes the version of the IP protocol, for example, version 4. The LEN field is the length of the IP *header* counted in 32-bit units. The type-of-service field describes the quality of service requested by the sender for this IP packet. It has the format

Precedence I D I T I R I xxx

where precedence is an indication of the priority of the IP packet; D specifies whether this IP packet can be delayed (0) or cannot be delayed (1); T indicates the type of throughput desired (0 = normal, 1 = high); R specifies whether reliable subnetwork is required (1) or not (0); and xxx is reserved for future use. The precedence options are routine (000); priority (001); immediate (010); flash (011); flash override (100); critical (101); internetwork control (110); and network control (111).

The total-length field specifies the length of the entire IP packet. Since the IP packet is encapsulated in the underlying network frame (e.g., LLC and then MAC), its length is constrained by the frame size of the underlying network. For example, as mentioned, the Ethernet limitation is 1500 octets. However, IP itself deals with this limitation

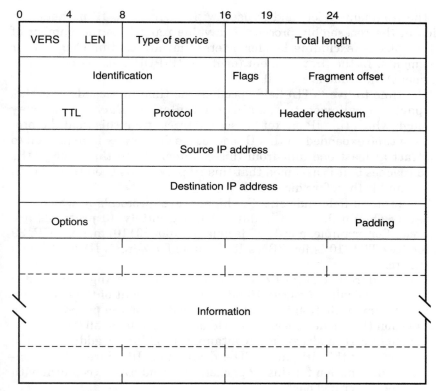

Figure 4.14 IP PDU.

by using segmentation and reassembly (SAR) (also called *fragmentation and defragmentation*). IP does require, however, that all underlying networks be able to handle IP packets up to 576 octets in length without having to use SAR capabilities. Fragments of an IP packet all have a header, basically copied from the original IP packet, and segments of the data. They are treated as normal IP packets while being transported to the destination. However, if one of the fragments gets lost, the entire IP packet is declared lost since IP does not support an acknowledgment mechanism; any fragments which have been delivered will be discarded by the destination.[4] More information on segmentation is provided below.

The identification field contains a unique number assigned by the sender to aid in reassembling a fragmented IP packet (all fragments have the same identification number). The flags field is of the form 0 | DF | MF, where DF specifies if the IP packet can be segmented (0) or not (1); and MF specifies if there are more segments (1) or no more segments, the present one being the last (0).

The fragment-offset field is used with fragmented IP packets and aids in the reassembly process. The value represents the number of 64-bit blocks (excluding header octets) that are contained in earlier fragments. In the first segment (or if the IP PDU consists of a single segment) the value is set to 0.

The time to live (TTL) field specifies the time in seconds that this IP packet is allowed to remain in circulation. Each IP gateway through which this IP packet passes subtracts from this field the processing time expended on this IP packet (each gateway is requested to subtract at least one unit from this counter). When the value of the field reaches 0, it is assumed that this IP packet has been traveling in a loop and is therefore discarded.

The protocol field indicates the higher level protocols to which this gateway should deliver the data. Approximately 40 protocols are defined. For example, a code of decimal 6 (=00000110) means TCP; 29 is for ISO TP4; 10 is for BBN's RCC; 22 is for Xerox's IDP; 66 MIT's RVD; etc.

The header-checksum field is a checksum covering the header (only). It is calculated as the 16-bit ones complement of the ones complement sum of all 16-bit words in the header (for the purpose of the calculation the header-checksum field is assumed to be all 0s).

The source IP address field contains the 32-bit IP address of the device sending this IP packet. The destination IP address field contains the destination for this IP packet. These addresses conform with the format described earlier.

The options field (which must be processed by all devices in the interconnected network, although not all devices must be able to generate such field) defines additional specific capabilities. These include explicit routing information, record route traveled, and timestamping.

4.4.6 IP segmentation

As IP packets travel from one network to another, it may run across networks that have a maximum frame size smaller than the length of the IP packet. This precludes the placement of an IP packet into a single network frame. A procedure is required to fragment long IP packets and reassemble them at the receiving end. An unfragmented IP packet has the "more segments" flag bit set to zero; the fragment offset is also set to zero. A gateway is not required to reassemble a fragmented packet, unless the packet carries the address of that gateway.

When fragmentation is required, the following tasks are undertaken.

1. The DF flag is checked to determine if fragmentation is allowed. If it is not allowed, the IP packet is discarded.

2. The information field is segmented into two or more parts, each having a length, in octets, that is a multiple of 8 (the last fragment is padded, if needed, to meet this criterion).
3. All information portions are placed in the IP packet. The header of the "continuation of message" packets is the same as the "begin of message" packet, with the following modifications:
    a. The "more fragments" flag is set to 1, except for the last fragment.
    b. The "fragment offset" field is set to the location this data portion occupied in the original IP packet, relative to the beginning of the original unfragmented IP packet.
    c. If options were included in the original IP packet, the type of option determines whether or not these options will be copied into all fragment IP packets or just to the first one.
    d. The header checksum field is recomputed.
4. Each of these fragmented IP packets is now transmitted as a normal IP packet.

At the destination, the information has to be reassembled into one IP packet. The "identification field" of the IP packet was set by the sending device as a unique number, from that sending device's perspective. The fragmentation process does not alter this field. Incoming fragments can, therefore, be identified and associated. (This ID must be used in conjunction with the source and destination IP address in the IP packet.) A buffer, managed by a timer, is employed at the destination to undertake the reassembly process.

4.4.7   Internet control message protocol

The ICMP can be used to support some error reporting in gateway with partial information. ICMP uses IP as if it were a higher-level protocol. However, ICMP is actually part of IP. It does not improve the reliability of IP: packets may still be undelivered without any report on their loss. (Reliability must in fact be implemented by the higher-level protocols, namely TCP.) ICMP can report on data packet loss, but not on ICMP message loss, to avoid infinite repetitions. For fragmented IP packets, ICMP messages are only sent about errors on the first fragment (the one with the 0 fragment offset field).

4.4.8   Address resolution protocol

The address resolution protocol (ARP) performs a key function: it is responsible for binding higher-level addresses (IP addresses) to physical network addresses. Individual hosts or devices are identified on the network by their physical hardware address.

As discussed, higher-layer processes (and protocols) address destinations using a symbolic IP address. When such a higher-layer process needs to send a packet to a destination, it uses its IP address expressed as w.x.y.z discussed earlier. The device driver does not know how to interpret this address. Consequently, the ARP module is provided to take care of the translation of the IP address to the physical address of the destination device. ARP uses a table (also known as the ARP cache) to perform this translation.

When the address is not found in the ARP cache, a broadcast message, called the *ARP request,* is sent over the network. If one of the devices on the network recognizes its own IP address in the request, it sends an *ARP reply* to the requesting entity. The reply contains the physical hardware address of the destination. This newly received address is then placed in the ARP cache of the requesting device. All subsequent packets to this destination IP address can now be directly translated to a physical address, which is used by the sender's device driver to transmit the packet on the network.

4.4.9   Reverse address resolution protocol

Some network devices, such as diskless workstations, do not immediately know their own IP address when they are booted. To determine their own IP address, they must use a mechanism similar to ARP. In this case, the hardware address of the device is the know parameter and the IP address the required parameter. This protocol differs from ARP in that a "RARP server" must exist on the network which maintains a database mapping from hardware address to protocol address.[4]

4.4.10   Interior-exterior gateway protocols

An organization may have several physical networks that are interconnected with (IP) gateways. This arrangement can be considered an autonomous system (*system* used in the sense of collection of parts), supported by a single management authority. Autonomous systems use noncore gateways. Figure 4.4 depicts this situation. An autonomous system maintains information on reachability of devices on that system; specified gateways must forward this routing and reachability information to other autonomous systems. Gateways that support this function can either be labeled as *exterior* or *interior.* An exterior gateway belongs to a different autonomous system; an interior gateway belongs to the same autonomous system.

Exterior gateways communicate the appropriate routing information using the exterior gateway protocol (RFC 904). The exterior gateway can forward reachability information for subnetworks within its own autonomous system. The information is collected by the exterior gateway using the interior gateway protocol.

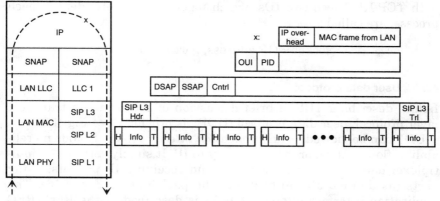

Figure 4.15 Example of router interconnection (for SMDS). PID = protocol ID; OUI = organizationally unique identifier; SAP = service access point; SSAP = source SAP; DSAP = destination SAP; SIP = SMDS interface protocol; H = header; T = trailer.

4.4.11 Routing over an SMDS network

A protocol suite for routing over SMDS has been developed (*RFC 1209: The Transmission of IP Datagrams over the SMDS Service*, March 1991). Figure 4.15 depicts the router operation in this context. The router matches the LAN protocols at one end and the SMDS protocols at the other end. PCs build TCP/IP packets destined for a remote destination. The IP address of the destination is appended by the PC before the PDU is enveloped into the SNAP, LLC, and MAC frames. At the MAC layer, the address of the local router is included. The router strips out the MAC (with its address), the LLC, and the SNAP overhead, exposing the IP address. The routing tables in the router indicate that the IP address corresponds to a remote location. The IP packet is then encapsulated into SNAP, LLC, and SMDS frame. SMDS utilizes the 60-bit SMDS address of the remote router. The SMDS network delivers the frame to the remote router. The inverse process takes place at the remote router to deliver the data to the appropriate user.

4.5 TCP

4.5.1 Ports and sockets

Each process that wants to communicate with another process (such as SMTP, FTP, TELNET, etc.) identifies itself to the TCP/IP protocol suite by one or more *ports*. A port, in this context, is a 16-bit ID used by the host-host protocol to identify to which higher-level protocol or application program it must deliver incoming messages.

A TCP/IP process is identified by (1) the IP address of the device on which it runs, and (2) the port number via which it communicates with TCP/IP. These two IDs, which together uniquely identify each process, are called a *socket*. Hence,

$$\text{socket} = (\text{IP address, port number}).$$

4.5.2 User data protocol

Before describing TCP, a brief discussion of a companion transport layer protocol, user data protocol (UDP), is provided (see Fig. 4.2).

UDP can be viewed as an *application interface to IP*. It adds no reliability, flow control, or error recovery to IP. It simply serves as a multiplexer/demultiplexer for sending and receiving IP packets, using ports (as defined above) to direct the packets (see Fig. 4.16). The application interface offered by UDP is described in the RFC (RFC 768: *User Datagram Protocol*). A UDP PDU consists of five fields:

1. Source port (16 bits)

2. Destination port (16 bits)

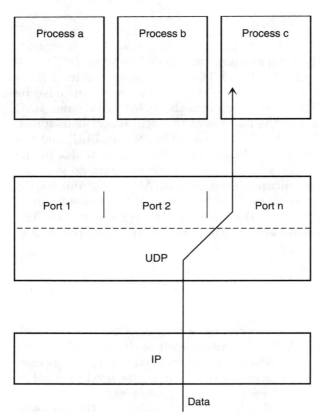

Figure 4.16 UDP function.

3. Length of packet, including header (16 bits)

4. Checksum (16 bits)

5. Data

Typical applications using UDP include:

- Trivial file transfer protocol (TFTP)
- Domain name server
- Remote procedure call (RPC), used by Network File System
- Simple network management protocol (SNMP)

4.5.3 Transmission control protocol

IP is an "unreliable," best-effort connectionless network layer protocol. This section discusses transmission control protocol (a transport layer protocol) that provides reliability, flow control, and error recovery. TCP is a connection-oriented, end-to-end reliable protocol providing logical connections between pairs of processes. Each process in a TCP/IP environment is uniquely identified by a socket. Ports are used for demultiplexing incoming data to the appropriate higher-level processes, as was the case in UDP. In TCP, a *connection* is uniquely defined by a *pair of sockets*; in other words, TCP is defined by a pair of processes, on the same or a different system, that are exchanging information.

The primary purpose of TCP is to provide reliable logical connections between pairs of processes, as shown in Fig. 4.17. Some TCP features are:[4]

- *Data transfer.*   From the applications viewpoint, TCP transfers a contiguous stream of octets through the interconnected network. The application does not have to segment the data into blocks or packets since TCP does this by grouping the octets in *TCP segments,* which are then passed to IP for transmission to the destination. TCP determines how to segment the data, and it forwards the data at its own convenience.

- *Reliability.*   TCP assigns a sequence number to each TCP segment transmitted and expects a positive acknowledgment (ACK) from the receiving peer TCP.* If the ACK is not received within a specified time-out interval, the segment is retransmitted. The receiving TCP uses the sequence numbers to rearrange the segments if they arrive out of order and to discard duplicate segments.

---

*In actuality, sequence numbers are assigned to each octet in the stream, but only the sequence number of the first octet in the stream is transmitted remotely.

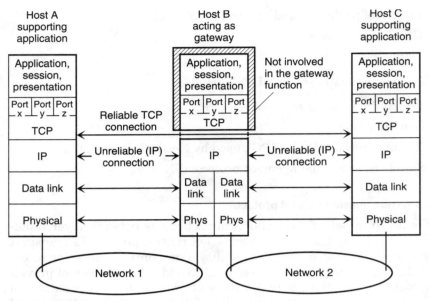

Figure 4.17   TCP reliability function.

- *Flow control.*   The receiving TCP signals to the sender the number of octets it can receive beyond the last received TCP segment, without causing an overflow in its internal buffers. This indication is sent in the ACK in the form of the highest sequence number it can receive without problems (this approach is also known as a window mechanism). It will be covered in more detail later in this chapter.

- *Multiplexing.*   Achieved through the use of the ports mechanism, as shown in Fig. 4.16.

- *Logical connections.*   In order to achieve reliability and flow control, TCP must maintain certain status information for each "data stream." The combination of this status, including sockets, sequence numbers and window sizes, is called a logical connection (also known as virtual circuit).

- *Duplex communication.*   TCP provides for concurrent data streams in both directions.

4.5.4   Protocol windowing

A transport protocol could send a PDU and then wait for an acknowledgment from the receiver before sending the next PDU. If the ACK were not received within a certain amount of time, the sending protocol entity could retransmit the PDU. While this mechanism would ensure reliability, it would only use a part of the available network bandwidth: the actual throughput of the link would be low because the sender would have to wait for twice the propagation delay, plus remote processing, before sending the next PDU. This is a particular problem when the delay (propagation and/or remote processing) is high. For example, consider sending a block of 1024 octets over a satellite-based T1 channel. The 8192 bits can be transmitted in 5.3 ms; the sender would have to wait 250 ms before the next block can be sent. This implies that only $8192 \cdot 4 = 32,768$ bits would be sent in 1 s, for an efficiency of 2 percent!

A windowing protocol aims at increasing the throughput. Assuming a window size K, such a protocol uses the following rules:

- The sender may send $n \leq K$ PDUs within the window without receiving an ACK, but must start a time-out timer for each of them.

- The receiver acknowledges each PDU received by indicating the sequence number of the last well-received PDU in a single ACK.

- The sender "slides the window" on each ACK received.

The windowing mechanism ensures (1) better throughput; (2) reliable transmission; and (3) flow control.

In TCP, sequence numbers are assigned to each octet in the stream. TCP partitions this stream into TCP segments to transmit them. Windowing is used at the octet level: the window size is expressed as a number of octets, rather than a number of PDUs. TCP segments transmitted and ACKs received include octet-sequence numbers. The window size is determined by the receiver and is variable during the data transfer. (Each ACK message includes the window size that the receiver is ready to deal with at that particular time.)

4.5.5   TCP segments

Figure 4.18 depicts the format of a TCP segment. Table 4.4 provides some information on the meaning of the fields.

4.5.6   Acknowledgments and retransmissions

Sequence numbers are counted octet by octet. Acknowledgments specify the sequence number of the next octet the receiver expects to receive.

4.5.7   TCP connections

Before any data can be transferred, a connection needs to be established between two processes. One of the processes issues a "passive

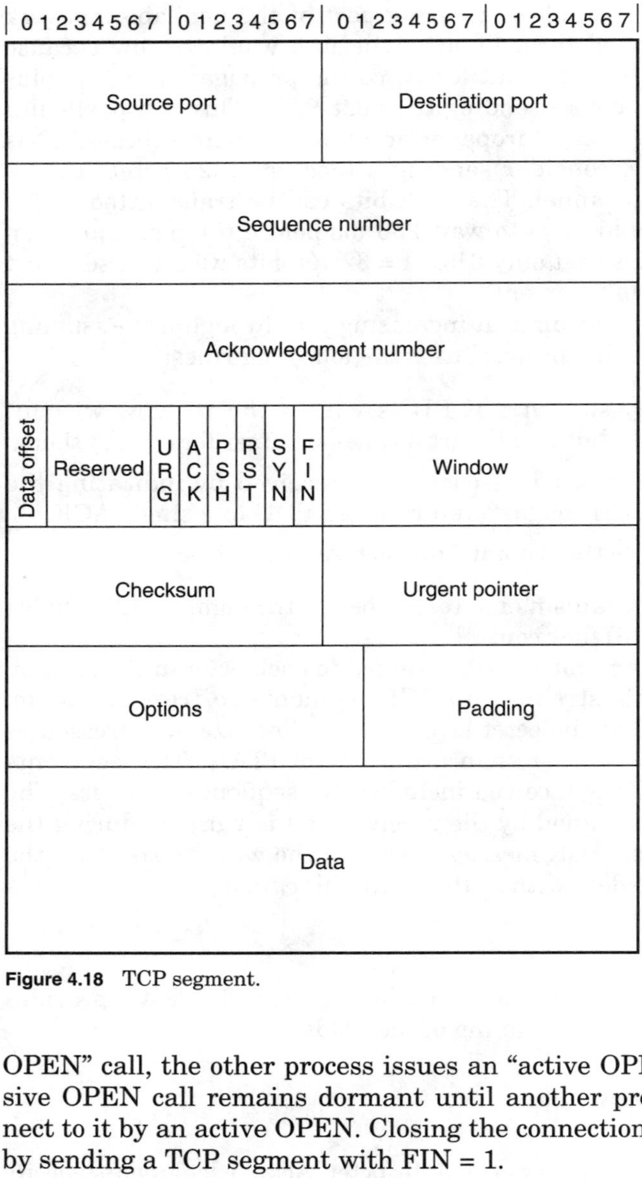

| 0 1 2 3 4 5 6 7 | 0 1 2 3 4 5 6 7 | 0 1 2 3 4 5 6 7 | 0 1 2 3 4 5 6 7 |

Figure 4.18   TCP segment.

OPEN" call, the other process issues an "active OPEN" call. The passive OPEN call remains dormant until another process tries to connect to it by an active OPEN. Closing the connection is done implicitly by sending a TCP segment with FIN = 1.

4.6   OSI LAN protocol suites

Figure 4.19 depicts a typical LAN protocol suite, when OSI protocols are used at the upper layers. (Note, however, that many commercially

TABLE 4.4   TCP Segment Fields

Source port	Port number used by the receiver to reply (16 bits).
Destination port	Destination port number (16 bits).
Sequence number	The sequence number of the first data octet in this segment. If the SYN control bit is set, the sequence number is the initial sequence number (n) and the first data octet is $n+1$ (32 bits).
Acknowledgment number	If the ACK control bit is set, this field contains the value of the next sequence number that the receiver device is expecting to receive (32 bits).
Data offset	The number of 32-bit words in the TCP header. It indicates where the data begins (4 bits).
Reserved	Reserved for future use; set to zero (6 bits).
URG	Indicates that the urgent pointer field is significant in this segment (1 bit).
ACK	Indicates that the acknowledgement field is significant in this segment (1 bit).
PSH	Push function (1 bit). Sometimes, an application needs to be sure that all the data passed to TCP has actually been transmitted to the destination. The push function is used for this purpose: it will "push" all remaining TCP segments still in storage to the destination.
RST	Resets the connection (1 bit).
SYN	Synchronizes the sequence numbers (1 bit).
Fin	No more data from sender (1 bit).
Window	Specifies the number of data octets beginning with the one indicated in the acknowledgment number field which the receiver is willing to accept. Used in ACK segments (16 bits).
Checksum	The 16-bit ones complement of the ones complement sum of all 16-bit words in a pseudo-header (see below), the TCP header and the TCP data. While computing the checksum, the checksum field itself is considered zero. The pseudo-header is comprised of the following fields (in the given order): source IP address; destination IP address; 00000000; protocol (= 6); and TCP length.
Urgent pointer	Points to the first data octet following the "urgent" data. Only significant when URG = 1.
Options	Either a single octet containing the option number or a variable length option.

Layer 7	FTAM, TP, VP, MHS, etc.
Layer 6	ISO 8822 and 8823 with ISO 8824 and 8825 (ASN.1/BER)
Layer 5	ISO 8326 and 8327
Layer 4	ISO 8072 and 8073 class 0, 1, 2, 3, or 4 as needed
Layer 3	Connectionless network protocol, ISO 8473
Layer 2	LLC, MAC: CSMA/CD, token ring, token bus
Layer 1	PMD: IEEE 802.3, .4, .5 (ISO 8802.3, .4, .5)

Figure 4.19   Typical OSI-based communications.

available LANs today do not actually use OSI-based communication at the upper layers.)[5]

4.6.1 TCP/IP-to-OSI migration

The transition from a TCP/IP environment to an OSI-based environment has been talked about by a multitude of academicians, consultants, seminar developers, and trade press writers over the past five years, giving rise to voluminous documentation. At the moment, it appears to this author that the initial enthusiasm of the late 1980s, when some of the OSI protocols finally made to standards (often, after a development period of five to eight years), has not translated itself into a plethora of business-grade hardware and software products. If at all, this transition in the ranks of the business community will occur later in the decade, possibly between 1995 and 1998.

This section provides a summary of the methods by which both non-OSI-to-OSI and TCP/IP-to-OSI transition strategies can be implemented.* Available approaches include the following:

1. *Encapsulation (also known as tunneling).* With encapsulation, protocol α operates "over the top" of protocol ß in the same layer. This is done by inserting the entire α PDU into the user data field of the ß PDU. This approach was briefly introduced in Chap. 2. An example of particular interest in terms of TCP/IP-to-OSI transition is encapsulating ISO 8473 CLNP (see Table 4.1) over IP, so that OSI-based end systems could continue to use the Internet. Conversely, one could encapsulate IP over ISO 8473, in order to carry PDUs generated by an existing end system (e.g., a workstation in a LAN) over an OSI-based network. Tunneling entails some inefficiencies because all PCI in two headers are retained. However, encapsulation has the advantage of being transparent in the sense that protocol ß treats protocol α as if it were normal data.

2. *Dual suites.* Providing end-system software so that it can run separate protocol suites, one for the existing protocol environment (TCP/IP) and the other for the new environment (OSI), is the most elementary approach to transition. Concerns with this approach include the system and management resources required to implement and maintain the two suites. In addition to cost factors, this approach could degrade quality of service (QOS) and throughput.

3. *Mixed suites.* This method aims at phasing in OSI protocols, starting with OSI layers 5, 6, and 7, while retaining the already

---

*This section is based on discussions with T. Bounman and E. W. Geer of Bell Communications Research.

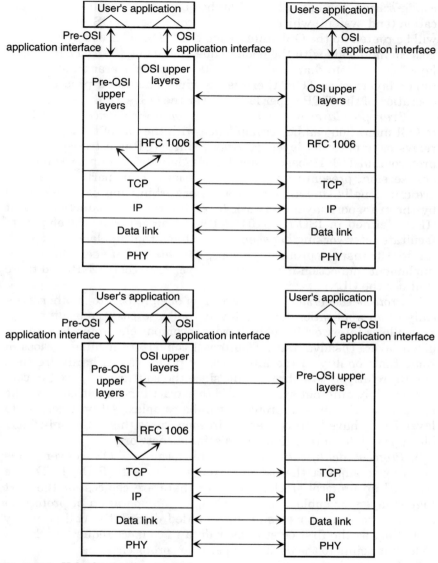

Figure 4.20 Hybrid suites.

installed networking capabilities of the TCP/IP suite (this is also known as hybrid suites). See Fig. 4.20. RFC 1006 describes how a TP0 OSI transport service, developed from the TCP data stream, can be used to provide a compatible interface to the OSI session layer. The PID mechanism, described below, is usually employed. With this configuration, the upper layers can be accessed as required: an application (end system) which has already converted to OSI upper layers will be routed up the OSI suite (using the PID); older end systems can also communicate with the node in question using the TCP/IP "upper-layer" suite. Note that while RFC 1006 supports operation of the OSI upper layers over TCP, there is no widely accepted procedure for operation of the "TCP upper layers" over the OSI's TP4.

4. *Transport layer relays and application layer gateways.* TCP/IP-to-OSI migration can be accomplished with the use of transport layer relays or application layer gateways, since TCP is a transport layer protocol and TCP/IP-based applications "bundle" the upper three layers (session, presentation, and application). Transport layer relays are functionally similar to the ISO interworking approach supported by the transport layer–network layer interworking functional unit (IFU) defined in ISO/IEC 10172-1991. The IFU was developed to facilitate interworking between end systems that implement dissimilar ISO transport protocol classes, as well as between those that implement connection-mode network layer protocol (CONP) and those that use the CLNP.

5. *Protocol conversion.* Conversion of one protocol to another, typically in an application layer gateway (but also possible at other layers), allows interworking of otherwise incompatible protocols within a given layer. Disadvantages of protocol converters include (1) loss of some functionality in the interworking protocols; (2) hardware cost; and (3) relatively low throughput, since the software required to convert possibly different syntaxes and to extract functions that have the same semantics in both protocols can be complex. All protocols up to layer 7 may have to be converted. In some cases this is the only feasible approach for interconnecting dissimilar networks.

6. *Protocol identification (PID).* This formalized OSI convention is one way to support the use of mixed stacks. With PIDs, PUDs are "routed" up protocol stacks whenever there are choices for the next higher layer (or sublayer). For example, the operation of protocol α over β described above could be signaled during the call setup by including, as the first octet of user data in connect request packet, a code that points to the desired upper layer protocol.

7. *Convergence functions.* Convergence functions are used to allow operation of adjacent protocol layers that would otherwise be incompatible. Typically, primitives are developed from a subnetwork service, or that service is manipulated to satisfy the underlying needs of the layer riding upon it. Use of RFC 1006, for example, allows operation of OSI session and higher layers over the TCP/IP stack in a "mixed-suites" approach. PID conventions are often included in the operation of convergence functions.

A brief discussion of tunneling follows.

4.6.2 Tunneling and network layer internetworking

In theory, network layer interworking provides more advantages and fewer disadvantages than the use of transport layer relays or application layer gateways. Hence, network layer interworking is the technically preferred method to achieve connectivity, according to observers. There are two forms of interworking within the network layer:

1. *End-to-end use of a connectionless-mode protocols.* "Tunnels" allow existing IP routers to "pass along" ISO 8473 packets on TCP/IP LANs. ISO 8473 packets are encapsulated inside IP packets, thereby allowing the "tunnel" to operate as a router. Figure 4.21 depicts how encapsulating can be used to carry CLNP over IP when IP is used as an internetworking protocol across an X.25-based WAN. Two convergence functions (CFs) are required: one for operation of CLPN/IP and another for IP/X.25. Work in this area has been proceeding in both the TCP/IP and OSI communities.

2. *Support of connection-oriented network layer service (CONS).* Early work on TCP/IP-to-OSI migration focused on OSI transport class 0 over TCP (using the approach of RFC 1006); this enabled OSI higher layers (session and above) to operate over a TCP/IP subnetwork. More recently, efforts have been directed to modify

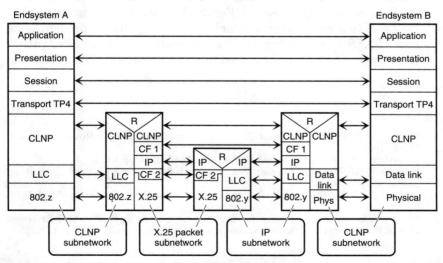

Figure 4.21 Tunneling with end-to-end CLNP. (ISO packets are encapsulated inside IP packets.) CF = convergence functions; R = relaying.

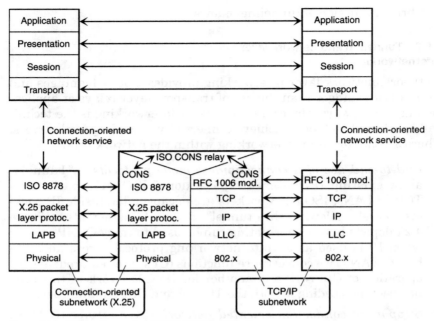

Figure 4.22 Tunneling using RFC 1006 (modified).

RFC 1006 so that the OSI CONS can be carried over TCP. CONS is the network layer services specified in ISO 8880-2, which is the companion *service* standard to the ISO network layer protocol over an X.25 network (ISO 8878). Figure 4.22 shows how interworking between two end systems is achieved using this approach. Note how CONS is operating as the internetworking service, rather than the ISO 8473 or IP connectionless protocols used in case (1) above.

References

1. D. Minoli, *Telecommunications Technology Handbook,* Artech House, Norwood, Mass., 1991.
2. B. Gerber, "IP Routing," *Network Computing,* April 1992, pp. 98 ff.
3. B. Gerber, "IP Routing: You Can Get Anywhere from Here," *Network Computing,* May 1992, pp. 134 ff.
4. IBM, *TCP/IP Tutorial and Technical Overview,* June 1990, Document GG24-3376-01.
5. Reston Consulting, "Strategies for Transitioning from TCP/IP to OSI Protocols Should Have a Practical Basis," *Networks In-Depth,* vol. 6, no. 3, 1992, p. 1.

Related LAN Standards

This chapter examines three relatively recent standards in support of traditional LANs:

1. Medium access control bridges: IEEE 802.1D
2. Simple network management protocols (SNMP)
3. Standards for Interoperable LAN Security (SILS): IEEE 802.10

5.1 Bridging Standards

5.1.1 Bridges

Bridges provide a means to physically extend the LAN environment and to provide improved performance. The bridged network of LANs (also referred to here as a *bridged system*) allows stations and servers attached to separate LANs to communicate as if they were attached to a single LAN. Bridges operate at the MAC layer. They provide frame-forwarding (relaying) functions within the MAC; each network retains its independent MAC, such as Ethernet or token ring, as shown in Fig. 5.1. Hence, a bridge, more precisely known as a "MAC bridge," interconnects the distinct LANs by relaying frames between the separate MAC sublayers.

As mentioned in Chap. 2, a MAC bridge operates below the MAC service boundary and is transparent to protocols operating above this boundary; this includes the LLC sublayer, the network layer (e.g., IP), the transport layer (e.g., TCP), and any other upper layer that may be employed by the user's application. Bridges support MA-UNITDATA.request primitives and corresponding MA-UNITDATA.indication primitives conveying user data with unconfirmed services.

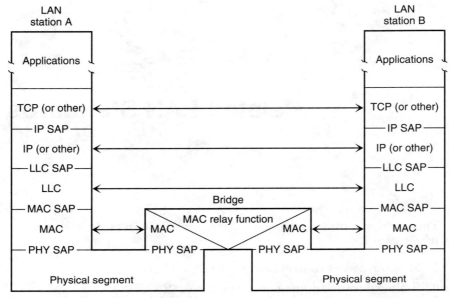

Figure 5.1 LAN bridge operation.

These are the basic elements of bridge operation:

1. Relay and filtering of frames
2. Maintenance of the information required to achieve frame filtering and relaying decisions
3. Management of these two functions

LAN bridges can be classified as pass-through and converting. Pass-through LAN bridges connect similar LANs and provide functions such as repeating, packet filtering, and packet queuing; frames are passed without any conversion. Converting bridges interconnect dissimilar LANs and must therefore convert the header and trailer of incoming frames (see Fig. 5.2).

Three typical corporate reasons for bridging LANs are as follows:[1]

1. They enable the interconnection of stations attached to 802 LANs of different MAC types (e.g., Ethernet and token ring).
2. They afford an increase in the physical extent, the number of permissible attachments, or the total performance of a LAN.
3. They facilitate partitioning of the physical LAN support for maintenance or administrative purposes.

The topic of *bridging* is being studied by the IEEE 802.1

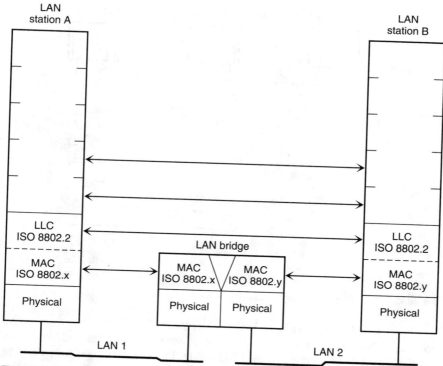

Figure 5.2 Pass-through and converting bridges. If x = y, then pass-through bridge; if x ≠ y, then converting bridge.

Interworking Task Group and by the IEEE 802.6 Bridging Subworking Group. There are two active aspects of investigation (see Fig. 5.3).[2]

- *Local bridging.* This covers environments where colocated bridges are directly connected, conforming to the 802.1D 1990 standard on transparent bridging. Although the two bridged LANs might have different MAC types, the same address structure is assumed. Hence, the address of an incoming frame is used or inserted directly into the (possibly different) outgoing frame.

- *Remote bridging.* This covers environments where the bridges are connected via an interconnecting medium, such as a T1 line, frame relay, or some other network. Typically, addressing resolution issues arise in this context. For example, connectivity by SMDS bridging falls in this category. SMDS, introduced in Chap. 1 and discussed more in Chap. 7, has an ITU-T E.164 60-bit address, while LANs have a 48-bit address. The issue of remote bridging over SMDS is being studied by the IEEE 802.6 Bridging Subworking Group (work is ongoing).

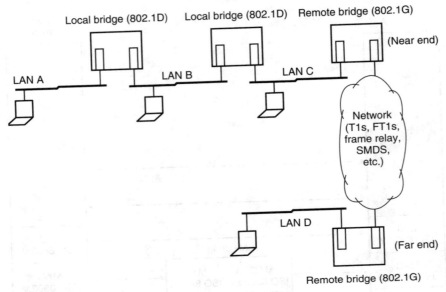

Figure 5.3 IEEE 802.1 bridging arrangements.

Figure 5.4 depicts a remote bridge operation. The MAC frame from the local LAN destined for a remote location is enveloped by the 802 SNAP. In turn, this PDU is enveloped by LLC fields. Finally, the resulting PDU is encapsulated in an 802.6 MAC frame for transmission. In this context, the SNAP uses the organizationally unique identifier (24-bit number taken from the IEEE 802 address space and used to identify an enterprise) and the protocol ID (to identify the protocol of the bridged MAC frame).

5.1.2 802.1D Bridges

A standard is now available defining an architecture for the interconnection of IEEE 802 LANs. The standard is known as IEEE 802.1D. The interconnection is achieved below the MAC service (i.e., layer) and is therefore transparent to the LLC and all upper layers. The standard includes a spanning-tree algorithm, which ensures a loop-free topology while at the same time providing redundancy. A 802.1D bridge operation is such that it provides redundant paths between end stations to enable the bridged LANs to continue to provide the service in the event of component failure of bridge or LAN. Figure 5.5 depicts an example of a set of interconnected LANs, with an eye to the spanning-tree algorithm discussed later.

Features supported by IEEE 802.1D bridges include the following:

- A bridge is not directly addressed by communicating end stations. Frames transmitted between end stations carry the MAC address

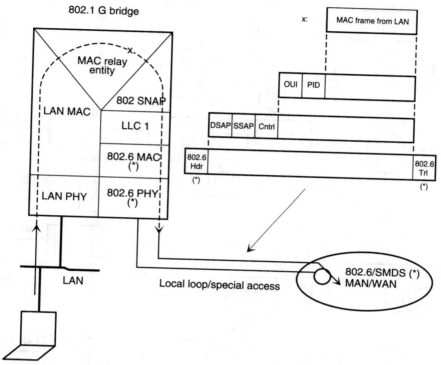

Figure 5.4 Remote bridging, IEEE 802.1G. PID = protocol ID; OUI = organizationally unique identifier; SAP = service access point; SSAP = source SAP; DSAP = destination SAP. (*) *Note:* For an SMDS network, the bridge implements the SMDS interface protocol (SIP) level 3, SIP level 2, and SIP level 1 in place of the 802.6 MAC and the 802.6 PHY. The header/trailer then conforms to the SIP level 3 protocol.

of the peer-end station in their destination address field, not the MAC address of the bridge (if it has any). The only time a bridge is directly addressed is as an end station for management purposes.

- All MAC addresses must be unique and addressable within the bridged system.
- The MAC addresses of end stations are not restricted by the topology and configuration of the bridged system.
- The quality of the MAC service (key aspects of which are shown in Table 5.1) supported by a bridge is comparable, by design, to that provided by a single LAN.

Service availability. As usual, availability is measured as that fraction of the (reference observation) time during which the service is provided in an unimpeded fashion. The availability is affected by the addi-

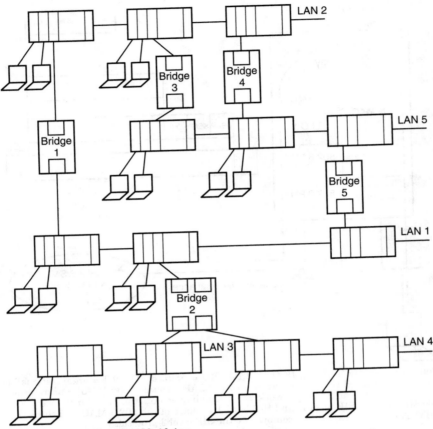

Figure 5.5 Example of LAN bridging.

TABLE 5.1   Factors of Interest in Bridging

Service availability
Frame loss
Frame misordering
Frame duplication
Transit delay experienced by frames
Frame lifetime
Undetected frame error rate
Maximum service data unit size supported
User priority
Throughput

tion of a component in a system. In the current case, the availability can actually be increased or decreased by the insertion of a bridge. The automatic reconfiguration around a failed link or component can clearly improve availability. However, unavailability of frame-forwarding resources (e.g., buffers), resulting from congestion and bridge failure, reduces the availability. The IEEE 802.1D standard aims at maximizing availability. Hence, it limits loss of service or delay to situations beyond its immediate control, such as failure, removal, or insertion of a component or bridge or movement on an end station to another segment of the interconnected system.

Frame loss. The service provided by the MAC sublayer does not guarantee the delivery of SDUs; however, frames transmitted by a source station arrive uncorrupted at the destination station with high probability. The operation of an IEEE 802.1D bridge introduces minimal additional frame loss.

In a bridged environment, a frame transmitted by a source station can fail to reach its destination station (in addition to obvious frame corruption due to noise during physical-layer transmission or reception) as a result of frame discard by a bridge. Discard can take place for the following reasons:[1]

- The bridge is unable to transmit them within some specified maximum time and, hence, must discard the frame to prevent maximum frame lifetime being exceeded.

- The bridge is unable to continue to store the frame owing to exhaustion of internal buffering as frames continue to arrive at a rate in excess of that at which they can be transmitted.

- The size of the SDU carried by the frame exceeds the maximum supported by the MAC procedures employed on the LAN to which the frame is to be relayed.

- Changes in the connected topology of the bridged LAN necessitate frame discard for a limited period of time to maintain other aspects of quality of service.

Frame misordering. The operation of the bridges does not misorder frames transmitted with the same user priority. However, where bridges in a bridged system are capable of connecting the individual MACs in such a way that multiple paths between source and destination station pairs exist, the operation of a protocol is required to ensure that multiple paths do not occur, since this situation can lead to misordering.

Frame duplication.   The service provided by the MAC sublayer does not permit the duplication of frames. The operation of a bridge does not itself introduce duplications of user data frames. However, the potential for frame duplication in a bridged system arises through the possibility of duplication of received frames on subsequent transmission within a bridge or through the possibility of multiple paths between source and destination end stations.

Transit delay.   The service provided by the MAC sublayer introduces a frame transit delay that is dependent on the particular medium and MAC method employed. Frame transit delay is the elapsed time between a MA-UNITDATA.request primitive and the corresponding MA-UNITDATA.indication primitive. Elapsed time values are calculated only on SDUs that are successfully transferred.

The *minimum* additional transit delay introduced by a bridge is the time required to receive a frame plus the time required to access the medium over which the frame is to be relayed. The frame is completely received before it is relayed since the FCS must be calculated and the frame discarded if in error. There can also be additional processing delays, making the delay greater than the stated minimum.

Frame lifetime.   The service provided by the MAC sublayer ensures that there is an upper limit to the transit delay experienced for a given instance of communication. This maximum frame lifetime is necessary to ensure the proper operation of higher-layer protocols. As discussed above, the bridge introduces additional transit delay. To enforce the maximum frame lifetime, a bridge may have to discard frames. Since the information provided by the MAC sublayer to a bridge does not include the transit delay already experienced by any particular frame, bridges must discard frames to enforce a maximum delay *in each bridge*. The value of the *maximum bridge transit delay* is based on both the maximum delays imposed by all the bridges in the bridged system and the desired maximum frame lifetime. A value of 1 s is recommended by the standard.

Undetected frame error rate.   The service provided by a LAN MAC sublayer introduces a very low undetected frame error rate in transmitted frames. Errors are identified by the use of an FCS. The FCS is appended by the sender and is checked by the destination station. The FCS calculated for a given service data unit is dependent on the MAC employed (token ring or Ethernet). It is therefore necessary to recalculate the FCS within a bridge providing a relay function between IEEE 802 MACs of dissimilar types (i.e., converting bridges described earlier). This introduces the possibility of additional undetected errors arising from the operation of a bridge, since the error rates are now additive. For frames relayed between LANs of the same MAC type, the 802.1D bridge does not introduce an undetected frame error rate greater than that which would be achieved by preserving the FCS across the system (i.e., computing it end to end).

5.1.3 Operation of an IEEE 802.1D bridge

This section provides information on the operation of an 802.1D bridge.[1] As indicated earlier, these functions include relaying and filtering of frames.

Relaying. An 802.1D bridge relays individual MAC user data frames between the separate MACs of the LANs connected to its ports. The order of frames of given user priority received on one bridge port and transmitted on another port is preserved. The functions that support the relaying of frames are shown in Table 5.2.

Filtering. A bridge filters frames in order to prevent unnecessary duplication. For example, frames received at a port are not copied to the same port. Frames transmitted between a pair of end stations are confined to LANs that form a path between those end stations. The functions that support the use and maintenance of filtering information include:[1]

- Automatic learning of dynamic filtering information through observation of bridged system traffic
- Aging of filtering information that has been automatically learned
- Calculation and configuration of bridged system topology

TABLE 5.2   Relaying Functions of a Bridge

Frame reception

Discard of frames received in error

Frame discard if the frame type is not user data frame

Frame discard following the application of filtering information

Frame discard if service data unit size exceeded

Forwarding of received frames to other bridge ports

Frame discard to ensure that a maximum bridge transit delay is not exceeded

Selection of outbound access priority

Mapping of service data units and recalculation of frame check sequence

Frame transmission

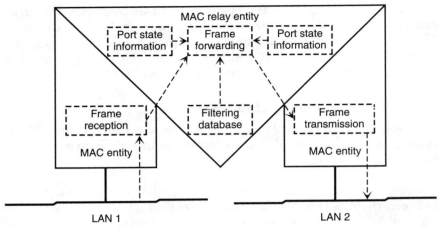

Figure 5.6 Internals of a bridge.

Bridge internals. Functionally, the bridge encompasses two entities (see Fig. 5.6): the MAC entity (the ports) and the MAC relay entity.

The *MAC entity* (bridge port) receives and transmits frames to and from the LAN to which it is attached. The MAC entity for each port handles the MAC protocol and procedures as specified in the IEEE 802 standard for that MAC technology. The key functions are as follows:

- *Frame reception.* The individual MAC entity associated with each bridge port examines all frames transmitted on the LAN to which it is attached. Frames that are in error as defined by the relevant medium access method are discarded; this includes frames whose FCS is in error.

- *Frame transmission.* The individual MAC entity associated with each bridge port transmits frames submitted to it by the MAC relay entity. Frames that are in error as defined by the relevant medium access method are discarded; this includes frames whose FCS is in error.

The *MAC relay entity* handles the functions of relaying frames between bridge ports, filtering frames, and learning filtering information. The key functions are:

- *Maintenance of port state information.* State information associated with each bridge port governs its participation in the bridged system. If management permits a port to participate in frame relay, and if it is capable of doing so, then it is described as active. The 802.1D standard specifies the use of a spanning-tree algorithm and protocol, which reduces the topology of the bridged system to a simply connected active topology.

- *Frame forwarding.* The forwarding process forwards received frames that are to be relayed to other bridge ports, filtering frames on the basis of information contained in the filtering database and on the state of the bridge ports, as shown in Fig. 5.6.

5.1.4   The spanning-tree algorithm and protocol

The configuration algorithm and protocol used in IEEE 802.1D reduce the bridged LAN topology to a single spanning tree. This pares down the complexity of the forwarding process. The spanning-tree algorithm and its associated bridge protocol operate to support, preserve, and maintain the quality of the MAC service. In order to perform this function, the algorithm performs the following tasks:

1. It configures the active topology of a bridged system of arbitrary topology into a single spanning tree, such that there is at most one data route between any two end stations, thereby eliminating loops.

2. It provides for fault tolerance by automatic reconfiguration of the spanning-tree topology in case of bridge failure or a breakdown in a data path, within the confines of the available network components, and for the automatic accommodation of any bridge port added to the bridged system without the formation of transient data loops.

3. It allows for the entire active topology to become stable (with a high probability) within a short, known interval. This minimizes the time for which the service is unavailable between any pair of end stations.

4. It allows for the active topology to be predictable and reproducible. This allows the application of configuration management, following traffic analysis, to meet the goals of performance management.

5. It operates transparently to the end stations; the stations are unaware of their attachment to a single LAN or a bridged system.

6. It keeps the communication bandwidth consumed by the bridges in establishing and maintaining the spanning tree to a small percentage of the total available bandwidth. Such bandwidth is independent of the total traffic supported by the bridged system regardless of the total number of bridges or LANs.

7. It keeps the memory requirements associated with each bridge port independent of the number of bridges and LANs in the bridged system.

8. It operates so that bridges do not have to be individually configured before being added to the bridged system, other than having their MAC addresses assigned through normal procedures.

The spanning-tree algorithm and protocol configure a simply connected active topology from the arbitrarily connected components of a bridged system. Frames are forwarded through some of the bridge ports in the bridged system and not through others, which are held in a blocking state. At any time, bridges effectively connect just the LANs to which ports in a forwarding state are attached. Frames are forwarded in both directions through bridge ports that are in a forwarding state. Ports that are in a blocking state do not forward frames in either direction but may be included in the active topology, i.e., be put into a forwarding state if components fail, are removed, or are added.

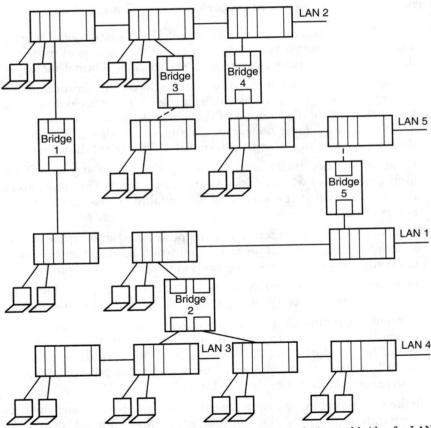

Figure 5.7 Example of designated bridges. Bridge 1 (root): designated bridge for LAN 1 and LAN 2. Bridge 2: designated bridge for LAN 3 and LAN 4. Bridge 4: designated bridge for LAN 5.

Figure 5.5 shows an example of a bridged system. Although there are physical loops in that topology, the algorithm applied to such topology ensures that there are no logical loops. One of the bridges in the bridged system is known as the *root* or the *root bridge*. Each individual LAN has a bridge port connected to it that forwards frames from that LAN toward the root and forwards frames from the direction of the root onto that LAN. This port is known as the *designated port* for that LAN; the bridge which this port is part of is known as the *designated bridge* for the LAN. The root is the designated bridge for all the LANs to which it is connected. The ports on each bridge that are in a forwarding state are the root port (that closest to the root) and the designated ports (if there are any). Note that one cannot tell which is the root of a bridged system simply by looking at the physical topology.

In Fig. 5.7, bridge 1 has been selected as the root and is the designated bridge for LAN 1 and LAN 2. Bridge 2 is the designated bridge for LAN 3 and LAN 4, and bridge 4 is the designated bridge for LAN 5. This gives rise to an active topology and in turn a logical tree topology (shown in Fig. 5.8).

The stable active topology of a bridged LAN is determined by:

- The unique bridge identifiers associated with each bridge
- The path cost associated with each bridge port (see below)
- The port identifier associated with each bridge port

The bridge with the highest-priority bridge identifier is the root. (For convenience of calculation, this is the identifier with the lowest numerical value.) Every bridge port in the bridged system has a root path cost associated with it.[1] This is the sum of the path costs for each bridge port receiving frames forwarded from the root on the least-cost path to the bridge. The designated port for each LAN is the bridge port for which the value of the root path cost is the lowest (if two or more ports have the same value of root path cost, then first the bridge identifier of their bridges and then their port identifiers are used as tie breakers). Consequently, a single bridge port is selected as the designated port for each LAN. The same computation selects the root port of a bridge from among the bridge's own ports. The active topology of the bridged system is completely determined by this process.

A component of the bridge identifier of each bridge and the path cost and port identifier of each bridge port can be managed, thus allowing a LAN manager to select the active topology of the bridged system.

5.1.5 Commercial outlook

This section provides a terse view of some practical product issues.

A designer can employ (1) simple bridges, (2) complex bridges, (3) routers, and (4) a combination of bridges and routers. Note there is an

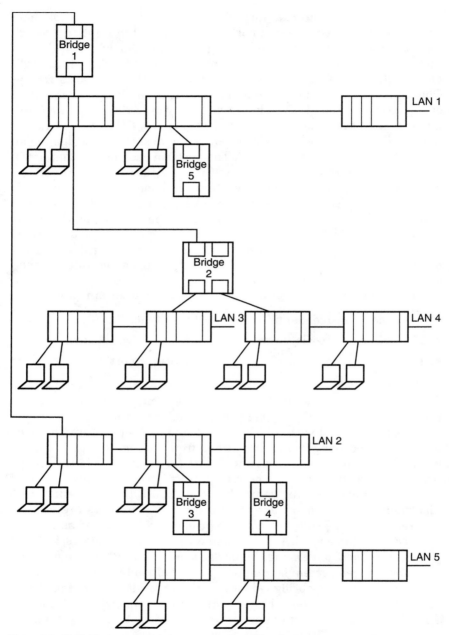

Figure 5.8 Logical view of spanning-tree configuration for previous example.

overlap in some of the functionality in actual commercial products, making the choice less clear-cut.

Basic bridges only provide frame filtering and forwarding functions. More sophisticated bridges support many other features (some described in Chap. 2), including the spanning-tree algorithm, which allows topology reconfiguration following a link failure. A survey of over 100 commercial bridges indicates that as of the date of publication about 65 percent of the products supported the spanning-tree algorithm.[3]

Bridges with more sophisticated filtering improve network management. Although the bridged system is more complex, it still only involves MAC-layer protocols, thereby ensuring protocol transparency immediately above the MAC layer. A multiport bridge may in fact cost more than a typical router. An advanced bridge generally costs 25 percent less than a typical router for the same number of ports. Also, a bridge is usually faster (higher filter-forward rate) than a router. This is because the additional processing of the network layer protocols can be resource-intensive. Bridges and routers typically employ the same microprocessor technology, but more machine cycles are required to deal with the network layer. Figure 5.9 depicts a histogram of 100 typical products in terms of frame-forwarding rates

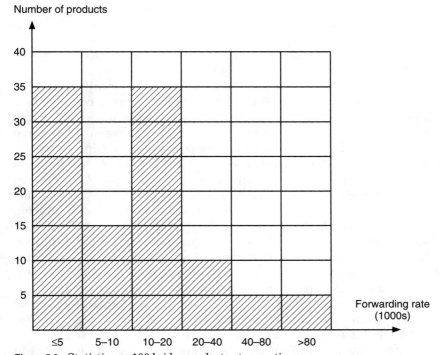

Figure 5.9 Statistics on 100 bridge products at press time.

Number of products

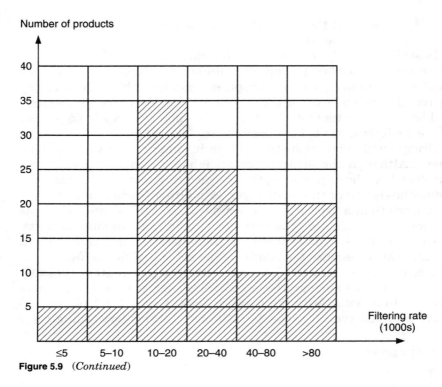

Figure 5.9 (*Continued*)

and filtering rates. Filtering determines if a frame is a local frame or a frame which needs to be forwarded; forwarding is the actual transmission of a frame to a downstream network. Typically, a bridge examines the source address of an incoming frame, constructs a source address table to associate it with the LAN on which it was received, and then looks for the destination address on the source address table. If the destination address is not found in the source address table, or if it is associated with another LAN, the frame is forwarded; otherwise (i.e., if there is a match) the packet is filtered and, hence, restricted to the LAN it is already on.[3] The "typical" forwarding rate is 10,000 to 20,000 frames/s and the "typical" filtering rate is 15,000 to 50,000 frames/s. Filtering rates are usually 2 to 3 times the forwarding rate. Stated filtering-forwarding rates may be difficult to interpret, since they depend on frame length; sometimes the information field has zero length. Even if "typical" frames are used, the numbers depend on the actual mix of short and long frames.

Figure 5.10 shows bridge costs at time of publication. Many of these products can be purchased for about $2000. Table 5.3 depicts a few high-end products, based partially on Cope.[3]

Filtering algorithms allow a determination of which frames need to be directed across the bridge to the downstream networks. Vendors

Number of products

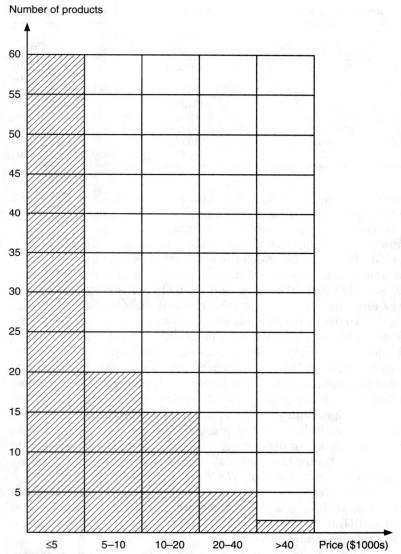

Figure 5.10 Cost statistics on 100 bridge products at press time.

have been adding more filtering capabilities to their bridge products, including user-defined filters. The more the filtering capabilities, the better the performance of the bridged system, since more effective utilization of bandwidth-bounded (WAN) links can be achieved. However, filtering requires bridge computing resources. A tradeoff between communication utilization and bridge computing power (equatable, ultimately, to delay) is required. When complex filtering slows down the bridge and degrades performance, a router may be preferred.

TABLE 5.3   A Few High-End Bridge Products Available at Press Time

Vendor	Product	Ports	Forwarding rate	Filtering rate	Price
Artel	Galactica	36	200,000	476,000	$20,000–$55,000
Fibronics	FX 8610	14	15,000	416,000	$32,000
Kalpana	EtherSwitch	15	104,000	223,000	$20,000
Penril	Series 2500	126	60,000	506,000	$44,000
Racal	RNX 6300	30	30,000	75,000	$18,000
Raycom	FddiRing	2	10,000	500,000	$19,000
3Com	ISOLAN	4	14,000	460,000	$15,000

As mentioned in Chap. 2, some vendors are combining the IEEE 802.1D transparent routing and the source-routing procedure in one device, described as source-routing transparent (see Fig. 2.21).

Routers provide more functionality than bridges, particularly with respect to WAN choices (of which there are many, as discussed in Chap. 7). Also, many LAN managers use bridges for local interconnection applications. However, they also require more network management resources and staff time. Some traffic is not directly routable (e.g., DEC's LAT). To deal with this situation, many routers have a bridging fallback mode to carry this traffic. The ability to dynamically support multiple network paths usually makes an interconnected network of routers (informally called an "internet"), more reliable. Additionally, since it can support multiple technologies, a route decision can be made on cost (least-cost routing) or grade of service (say, lowest delay). The spanning-tree algorithm enables a bridge implementing it to use a redundant path or link in case of primary link failure; however, since these links are not being utilized on a normal dynamic basis, as would be the case for a router (say, for load-sharing purposes), this topology may be expensive, particularly in WAN applications.

Many interconnected systems utilize both bridges and routers; the decision issues include cost and underlying traffic protocols. Besides access to multiple WAN services, routers offer flow control, even when used in a local application (see Fig. 5.11).

Given a certain user population and a specified level of performance, local interconnection of LANs using the same technology (e.g., Ethernet only or token ring only) is generally cost-effective with bridges. Given a certain user population and a specified level of performance, local interconnection of LANs using multiple technologies may ultimately be cheaper with routers (except, at this time, between LANs and an FDDI backbone, where bridges do better).

Each situation needs to be examined on its own merit, and no generalization is really sound.

Figure 5.11   NETBuilder II high-performance, multiprotocol bridge/router. (*Courtesy of 3Com Corporation.*)

5.2   Simple Network Management Protocol

Network management is an important aspect of corporate LAN operations.[4] In some cases the network management cost is as high as 40 percent of the total network cost (transmission and amortized equipment cost each typically being 30 percent of the total). The next chapter covers practical aspects of network management. This section covers some protocol-related issues.

5.2.1   Overview

The simple network management protocol (SNMP) RFC 1157 (May 1990) is a specification of a protocol for the management of interconnected networks. SNMP allows transfer of data pertaining to fault, performance, and configuration management of TCP/IP systems in general and LANs in particular. SNMP evolved from the simple gateway monitoring protocol (SGMP), which, in the late 1980s, was a mature, successful, and widely deployed Internet protocol. SGMP was changed based on operational experience to provide management (not just control) of facilities beyond gateways.

SNMP was declared a draft standard by the Internet Activities Board in March 1988. It was elevated to recommended standard status in April 1989, with the release of the common management protocol over TCP/IP (CMOT) specification. As of May 1990, SNMP became a full Internet standard and is known as RFC 1157. The desire to ease eventual transition to OSI-based management protocols led to the definition, in the Abstract Syntax Notation 1 (ASN.1) language, of a structure of management information (SMI-RFC 1155, May 1990, and earlier RFC 1065) and management information base (MIB-RFC 1156, May 1990, and earlier RFC 1066). A new MIB, known as MIB-II was defined in RFC 1213 (March 1991). (In the sequel the term MIB refers specifically to MIB II.)

As of 1993 hundreds of vendors have developed products based on SNMP. In the absence of interoperating and fielded OSI-based network management systems, SNMP has become a de facto standard for management of multivendor TCP/IP-based interconnected networks. SNMP agents support about 75 percent of the available bridge products and 80 percent of the routers. About 40 vendors (including HP, Sun, Novell, SynOptics, DEC, Cisco) offer what is known as a "generic SNMP manager." Such "generic" agents can query any SNMP agent (for all MIB-II objects; see below) and can import a legitimate "draft-status" or private MIB. Observers are expecting to see many low-end managers for PCs based on Microsoft Windows, rather than on expensive workstations.

RFC 1157 (a rerelease of RFC 1098) defines a simple protocol by which management information for a network element may be inspected or altered by logically remote users. In particular, together with its companion RFCs, which describe the structure of management information along with the management information base, the protocol provides a simple, workable architecture and system for managing TCP/IP-based interconnected networks. SMI describes how managed objects contained in the MIB are defined. The MIB is a virtual information store.

Table 5.4 enumerates the key RFC specifications in support of SNMP.[5]

The Internet Activities Board recommends that all IP and TCP implementations be network-manageable. This implies implementation of the MIB-II (RFC 1213) and the SNMP (RFC 1157). The SNMP protocol has five message types:

- GetRequest
- GetNextRequest
- GetResponse
- SetRequest
- Trap

Traps track events such as Cold Start, Warm Start, Link Down, Link Up, Authentication Failure, Neighbor Loss, as well as vendor-specific events.

TABLE 5.4   Key SNMP Specifications

SNMP	RFC 1157*
SMI	RFC 1155†
MIB II	RFC 1213‡
SNMP over IPX suites	RFC 1298
SNMP security protocols	RFC 1352
SNMP administrative model	RFC 1351
A convention for describing SNMP-based agents	RFC 1303
SNMP over OSI suites	RFC 1283
SNMP communication services	RFC 1270
SNMP MUX protocol and MIB	RFC 1227
SNMP over Ethernet	RFC 1089

*Earlier document: RFC 1098.
†Earlier document: RFC 1065.
‡Earlier documents: RFC 1066 and RFC 1156.

SMI is a policy statement describing how the managed objects (hosts, terminals, routers, bridges, gateways, etc.) are defined. Such definition basically involves agreement on a standard language for management; the notation employed is the ASN.1, mentioned above. One of the basic definitions in the SMI pertains to the representation of data elements.

Each object has a name, a syntax, and an encoding. The name also specifies the object type. The object type together with an object instance is employed to uniquely identify a specific instatiation of the object. The syntax defines the (abstract) data structure corresponding to the given object type. The encoding of the object type specifies how that object type is represented using the object's syntax. An example of an object is shown in Fig. 5.12, which also shows a vendor-specific trap.

The MIB represents the database of management information and parameters. The management parameters are stored in a tree-structured database. All mandatory variables listed there must be supported, either statically (with null or default value) or dynamically (i.e., actually employed by the device in question).[6]

MIB-II contains about 170 objects and, as of September 1991, renders obsolete the previous MIB definition (RFC 1156). There are "draft," "proposed standard," and "experimental" MIBs for about five dozen network and device types, including T1 networks, Ethernet, token ring, bridge, FDDI network, SMDS network, frame relay network, and T3 network. There are also private or vendor-specific MIBs (numbering over 500 at writing time).

SNMP provides a means of communicating information between the management station and the managed devices. An SNMP user-

```
Object:

logicalToExplicitHub OBJECT-TYPE

SYNTAX INTEGER (1..65335)

ACCESS read-write

STATUS optional

DESCRIPTION
        "The value of this object identifies the hub for which this entry contains management
        information."

::={logicalToExplicitEntry 2}


Trap:

preselectedVCChange TRAP-TYPE

ENTERPRISE minoliinc

VARIABLES {ifIndex}

DESCRIPTION
        "A preselectedVCChange trap signifies that the preselection of the signaling virtual
        channel has been changed."

::=7
```

Figure 5.12 Example of an object and a trap.

manager issues instructions (set, get, etc.) to managed objects; the SNMP agent executes the instructions and responds.

The IAB recently directed the Internet Engineering Task Force (IETF) to create two new working groups in the area of network management. One group was charged with the further specification and definition of elements to be included in the MIB. This resulted in the MIB-II. The other (SNMP Extensions Working Group) was charged with defining the modifications to the SNMP to accommodate the short-term needs of the network vendor and operations communities. This resulted in a new protocol, the simple management protocol (SMP), in draft form as of press time. SMP is discussed in a later section.

Although SNMP uses a subset of ASN.1 and ISO common management information protocol (CMIP) uses ASN.1, it does not follow that the SNMP protocol is a subset of the CMIP protocol. SNMP is not upward-compatible with CMIP. A two-stage approach for network management of TCP/IP-based interconnected networks was initially planned by IAB. In the short term, SNMP was to be used to manage nodes in the Internet community. In the long term, the use of the OSI network management framework was to be examined. Both RFC 1065 and RFC 1066 were designed so as to be compatible with both the SNMP and the OSI network management framework. *RFC 1109 notes that the requirements of the SNMP and the OSI network management frameworks were more different than anticipated. As such, the requirement for compatibility between the SMI/MIB and both frameworks was suspended.* This action permitted the operational network management framework, the SNMP, to respond to new operational needs in the Internet community by producing documents defining new MIB items.

5.2.2 SNMP structure

The modeled environment consists of a collection of network management stations and network elements, as illustrated in Fig. 5.13. Network management stations execute management applications that monitor and control network elements. Network elements are devices such as hosts, gateways, terminal servers, which have management agents responsible for performing the network management

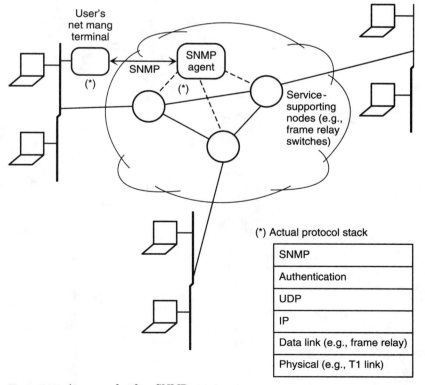

Figure 5.13 An example of an SNMP agent.

functions requested by the network management stations. SNMP is used to communicate management information between the network management stations and the agents in the network elements. The strategy implicit in the SNMP is that the monitoring of network state at any significant level of detail is accomplished primarily by polling for appropriate information by the monitoring center(s). A limited number of unsolicited messages guides the timing and focus of the polling. The SNMP aims at minimizing the number and complexity of

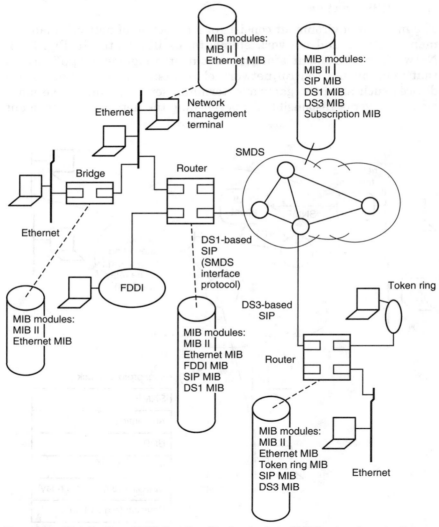

Figure 5.14   Distribution of MIP functionality (example in SMDS network context).

management functions realized by the management agent itself. With this approach, the development cost for the management agent software necessary to support the protocol is reduced. The degree of management function that is remotely supported is increased, while imposing the fewest possible restrictions on the form and sophistication of management tools. Figure 5.14 depicts an example of distributed functionality (modeled after Cox et al.[6]). The discussion which follows is based directly on RFC 1157.[7]

The SNMP architecture supports a wide range of administrative relationships among entities that participate in the protocol. The SNMP architecture supports a solution to the network management challenge in terms of the following items.

1. *Scope of the management information communicated by the protocol.* The scope of the management information communicated by operation of the SNMP is that represented by instances of all nonaggregate object types either defined in MIB or defined elsewhere according to the conventions set forth in SMI.

2. *Representation of the management information communicated by the protocol.* Management information communicated by operation of the SNMP is represented according to the subset of the ASN.1 that is specified for the definition of nonaggregate types in the SMI. A moderately complex subset of ASN.1 is used for describing managed objects and for describing the protocol data units used for managing those objects.

3. *Operations on management information supported by the protocol.* The SNMP models all management agent functions as "alterations" or "inspections" of variables. Thus, a protocol entity on a logically remote host interacts with the management agent resident on the network element in order to "get" or "set" variables. This has the effect of limiting the number of essential management functions realized by the management agent to two: one operation to assign a value to a specified configuration or other parameter and another to retrieve such a value. It also avoids introducing into the protocol support for imperative management commands. The exclusion of imperative commands from the set of explicitly supported management functions is not perceived as precluding any desirable management agent operation. Currently, most commands are requests either to set the value of some parameter or to retrieve such a value, and the function of the few imperative commands currently supported can be accommodated by this management model. For example, an imperative command might be realized as the setting of a parameter value that subsequently triggers the desired action.

4. *Form and meaning of exchanges among management entities.* The communication of management information among management

entities is realized in the SNMP through the exchange of protocol messages. Consistent with the goal of minimizing complexity of the management agent, the exchange of SNMP messages requires only an unreliable datagram service; every message is entirely and independently represented by a single transport datagram. SNMP specifies the exchange of messages via the UDP protocol introduced in Chap. 4; however, the mechanisms of the SNMP are generally suitable for use with a wide variety of transport services.

5. *Definition of administrative relationships among management entities.* *SNMP application entities* are entities residing at management stations and network elements which communicate with one another using the SNMP. *Protocol entities* are peer processes which implement the SNMP and thus support the SNMP application entities. An SNMP community is a pairing of an SNMP agent with some set of SNMP application entities. Each SNMP community is named by a string of octets, this being called the *community name.* An *authentic SNMP message* is a message originated by an SNMP application entity that belongs to the SNMP community named by the community component of said message.

An *authentication scheme* is the set of rules by which an SNMP message is identified as an authentic SNMP message for a particular SNMP community. An *authentication service* is an implementation of a function that identifies authentic SNMP messages according to one or more authentication schemes. Management of administrative relationships among SNMP application entities requires authentication services that are able to identify authentic SNMP messages with a high degree of certainty. (Some SNMP implementations may need to support only a trivial authentication service that identifies all SNMP messages as authentic SNMP messages.)

For any network element, an *SNMP MIB view* is a subset of objects in the MIB that pertain to that element. An element of the set {READ-ONLY, READ-WRITE} is called an *SNMP access mode.* An *SNMP community profile* is a pairing of a SNMP access mode with a SNMP MIB view. An SNMP community profile represents specified access privileges to variables in a specified MIB view. For every variable in the MIB view in a given SNMP community profile, access to that variable is represented by the profile according to the following conventions:

- If said variable is defined in the MIB with "Access:" of "none," it is unavailable as an operand for any operator.
- If said variable is defined in the MIB with "Access:" of read-write or write-only and the access mode of the given profile is READ-WRITE, that variable is available as an operand for the get, set, and trap operations.

- Otherwise, the variable is available as an operand for the get and trap operations.
- In those cases where a write-only variable is an operand used for the get or trap operations, the value given for the variable is implementation-specific.

An *SNMP access policy* is a pairing of an SNMP community with an SNMP community profile. An access policy represents a specified community profile made available by the SNMP agent of a specified SNMP community to other members of that community. For every SNMP access policy, if the network element on which the SNMP agent for the specified SNMP community resides is not that to which the MIB view for the specified profile pertains, then that policy is called an *SNMP proxy access policy*. The SNMP agent associated with a proxy access policy is called a *SNMP proxy agent*. Proxy policies permit the monitoring and control of network elements which are otherwise not addressable (a proxy agent may provide a protocol conversion function, allowing a management station to apply a consistent management framework to all network elements, including devices which support different management frameworks).

6. *Form and meaning of references to management information.* The SMI requires that the definition of a conformant management protocol address:

- The identification of particular instances of object types defined in the MIB. Each instance of any object type defined in the MIB is identified in SNMP operations by a unique name called its *variable name*. The name of an SNMP variable is an *object identifier* of the form x.y, where x is the name of a nonaggregate object type defined in the MIB and y is an object identifier fragment that identifies the desired instance.
- The resolution of ambiguous MIB references.
- The resolution of MIB references in the presence multiple MIB versions.

Table 5.5 summarizes some of these key concepts.

5.2.3 Protocol specification

The network management protocol is one by which the variables of an agent's MIB may be inspected or altered. Communication among protocol entities is accomplished by the exchange of messages, each of which is entirely represented within a single UDP datagram using the basic encoding rules (BER) of ASN.1.

A message consists of a version identifier, an SNMP community name, and a PDU. A protocol entity receives messages at UDP port 161 on the host with which it is associated for all messages except for

TABLE 5.5   Key SNMP Concepts: A Practical View

Agent	Software capability built into an SNMP-managed device. Key features: contains an IP address entity; knows the address of a manager to which traps are issued; supports a set of SNMP objects (from a few to several hundred types); and uses UDP/IP to transmit messages between managers and agents).
Community name	A security mechanism (password, or alphanumeric character string) included in each SNMP message. Features: predefined to the managed device; provides "weak" security; tiered hierarchical structure of managed objects is possible. *Secure SNMP* supports stronger security mechanisms (implementations should be available in 1993 and beyond).
MIB	A collection of objects (up to 200 types), structured as specified in RFC 1213. Private MIBs contain objects that only the agent software from a specified vendor supports.
Object	Formal description of an element of management information. Objects comprised of parameters such as object ID, object type, values (integer, counter, or string), and access (read-write, read-only, etc.).
Get message	The GetRequest message is issued by an SNMP management station to an agent. The GetResponse message contains response information from the device or agent back to the station. Either single objects or multiple objects can be queried or responded to by a single get message.
Set message	Message that enables operator at an SNMP management station to change the value of an object contained at an SNMP agent (as long as the object is defined as read-write). Used to control managed devices.
Trap message	Message issued unilaterally by a managed device upon recognition that a certain predefined condition or threshold has been met. Six trap conditions are defined in SNMP, but vendors have added vendor-defined traps. Of the specified SNMP traps, the Cold Start (power supply interruption) and the Authentication Failure (unauthorized attempt to access the device's agent) are the two most commonly implemented.

those which report traps (messages which report traps are received on UDP port 162). See Chap. 4 on the issue of UDP ports. An implementation of this protocol need not accept messages whose length exceeds 484 octets.

It is mandatory that all implementations of the SNMP support the five PDUs identified earlier: GetRequest-PDU, GetNextRequest-PDU, GetResponse-PDU, SetRequest-PDU, and Trap-PDU.

The top-level actions of a protocol entity that generates a message are as follows:

1. It first constructs the appropriate PDU as an ASN.1 object.

2. It then passes this ASN.1 object—along with a community name, its source transport, and the destination transport address—to the

service, which implements the desired authentication scheme. This authentication service returns another ASN.1 object.

3. The protocol entity then constructs an ASN.1 message object, using the community name and the resulting ASN.1 object.

4. This new ASN.1 object is then serialized, using BER/ASN.1, and then sent using a transport service to the peer protocol entity.

The top-level actions of a protocol entity that receives a message are as follows:

1. It performs a rudimentary parse of the incoming datagram to build an ASN.1 object corresponding to an ASN.1 message object. If the parse fails, it discards the datagram and performs no further actions.

2. It then verifies the version number of the SNMP message. If there is a mismatch, it discards the datagram and performs no further actions.

3. The protocol entity then passes the community name and user data found in the ASN.1 message object, along with the datagram's source and destination transport addresses, to the service, which implements the desired authentication scheme. This entity returns another ASN.1 object or signals an authentication failure. In the latter case, the protocol entity notes this failure, (possibly) generates a trap, and discards the datagram and performs no further actions.

4. The protocol entity then performs a rudimentary parse on the ASN.1 object returned from the authentication service to build an ASN.1 object corresponding to an ASN.1 PDU object. If the parse fails, it discards the datagram and performs no further actions. Otherwise, using the named SNMP community, the appropriate profile is selected, and the PDU is processed accordingly. If, as a result of this processing, a message is returned then the source transport address that the response message is sent from will be identical to the destination transport address to which the original request message was sent.

For detailed operation of the protocol, the reader is referred directly to RFC 1157.

5.2.4 Simple management protocol

A more "robust" version of SNMP was announced on July 15, 1992, at the twenty-fourth IETF meeting, along with accompanying draft specifications (see Table 5.6). The protocol is known as simple man-

TABLE 5.6   Eight 1992 Documents Describing SMP

Introduction to SMP

SMI for SMP

Textual Conventions for SMP

Protocol Operations for SMP

Transport Mapping for SMP

MIB for SMP

Manager to Manager MIB for SMP

SNMP/SMP Coexistence

Proposed by Case, McCloghrie, Rose, and Waldbusser.

agement protocol (SMP). At the time of publication it was expected that SMP would be adopted as a standard. SMP addresses some of the weaknesses of SNMP, including manager interaction (manager-to-manager communication), efficient file transfer from managed elements, and ability to run on a variety of protocol platforms, including AppleTalk, NetWare, and OSI (rather than being tied to UDP/IP). Table 5.7 depicts key differences. SMP is based on the secure SNMP. SMP is being viewed as an evolution from SNMP, since it can manage both SMP- and SNMP-equipped devices and applications. Observers and proponents expect a successful market penetration.[8]

5.3   LAN Security

Security is important to all networked systems, including LANs. The number of computer systems which are linked by communication networks and which require protection against various threats is increasing. While openness and high connectivity in a network are desirable goals, they give rise to security concerns.

Because of advances in LAN interconnection technology, the risks to LANs have become more severe. Campus LANs (e.g., FDDI), bridges and routers, communications gateways accessible from the

TABLE 5.7   Key Differences Compared to SNMP

Richer SMI (better able to create new objects, 64-bit counters, enumerated bitstrings, etc.)

Multiprotocol for transport

Number of error codes has been increased

New operator to ship files (GetBulk)

Can manage applications, not just devices

Secure

public network, fax servers, and other LAN server facilities have increased the range of resources which are vulnerable to abuse. Ring networks not only make every PDU (e.g., packet at the network layer or frame at the data link layer) available to every station on the LAN, but require every station to receive and then forward every data unit. These issues have prompted the concerns that lead to the development of Standards for Interoperable LAN Security (SILS).[9,10]

Security encompasses measures used to protect information from unauthorized disclosure, modification, or destruction. Security controls and mechanisms have been added to the basic OSIBRM to protect the exchanges of information (Addendum 2 of the OSIBRM (ISO 7498)). In addition to this security architecture, standards work is under way, and there are several national and international groups for standardizing approaches to information security for certain applications. New IEEE 802.10 standards specific to LANs are emerging.

The first few subsections focus on the OSIBRM security architecture. They describe the motivation for security. Some of the numerous threats faced by the network manager and covered by the standards are discussed. The services which these managers need in order to undertake security management tasks, which are described by the standards, are identified. This is followed by listing of available mechanisms for the management of security. Later subsections discuss how the evolving international standards have been applied to the LAN environment in the IEEE 802.10 standards.*

5.3.1 Motivations for security in open systems

In a computer system network environment, security refers to minimizing the risk of exposure of assets and resources to vulnerabilities and threats of various kinds. A vulnerability is any weakness of a system which could be exploited to circumvent the policy of the system. A threat is a potential violation of the security of the system.

5.3.2 OSI security architecture

There was an early recognition on the part of ISO Subcommittee SC21 that provision had to be made for security in the OSIBRM (ISO 7498). The OSI security architecture was the first of the SC21 security standards and was published in 1989 as Part 2 of the OSI basic reference model (ISO 7498-2). However, an architectural document

---

*Some nuances of what is specified in the IEEE 802.10 standard do not fit precisely within the OSI security architecture (key difference: OSI only defines confidentiality at the data link layer). However, other differences are small.

such as ISO 7498-2 is only the first stage in defining security services and mechanisms. Since it is not an implementation specification, additional standards are required that build on the concepts of the architecture. This has led to the development of related documents—security models, security frameworks, and proposals for specifying particular types of protection in the existing OSI protocol standards.

The OSI security architecture has two goals:

- It provides a general description of security services and related mechanisms, which may be provided by the OSI basic reference model.

- It defines the positions within the basic reference model where services and mechanisms may be provided.

Four key aspects are included in the OSI security architecture and are discussed in the sequel:

1. Security services

2. Security mechanisms

3. Appropriate placements of security services for all layers of the basic reference model

4. Architectural relationships of the security services and mechanisms to the basic reference model

ISO 7498-2 identifies five basic security services: access control, authentication, data confidentiality, data integrity, and nonrepudiation. These services provide protection against the security threats of unauthorized resource use, masquerade, unauthorized data disclosure, unauthorized data modification, and repudiation, respectively.

The OSI reference model security architecture is concerned only with the aspects of a communications path which permit open systems to achieve secure transfer of information between them. This type of security is not concerned with security measures needed within the systems themselves or the installation. The definition of security services to support such additional security measures is outside the scope of this standard. Hence, additional security measures are needed in end systems (e.g., gateways, servers, hosts, etc.), installations, and organizations.

5.3.3   Security threats

The threats which a network manager must consider, as identified in the OSI addendum are shown in Table 5.8.

TABLE 5.8   Key Security Threats

Denial of service	A form of attack where an entity acts so as to prevent other entities from performing their proper function
Masquerade	A type of attack where an entity asserts it is a different entity
Modification of messages	A threat that exists when the content of a data transmission is altered without detection and results in an unauthorized effect
Unauthorized disclosure	A threat whereby data which should be kept confidential is compromised by being made available to unauthorized agents
Replay	A threat that is carried out when a message, or part of a message, is repeated in order to produce an unauthorized effect
Trapdoor	A capability planted when an entity of the system is modified to allow an attacker to produce a future unauthorized effect on a command, or at a predetermined event
Trojan horse	An entity that, when innocuously introduced into the system, has a deliberately planned unauthorized effect in addition to its authorized function

5.3.4   OSI security services

This section describes the security services that are included in the OSI security architecture and the mechanisms that implement them in order to manage the security threats discussed above. The following basic services are defined:

- *Authentication.*   The authentication services provide for the verification of the identity of a remote communicating entity and the source of data.

- *Access control.*   The access control service provides protection against unauthorized use of resources accessible via the network.

- *Data confidentiality.*   These services provide for the protection of data from unauthorized disclosure.

- *Data integrity.*   This service provides for the integrity of all user data or of some selected fields over a connection or connectionless message exchange; it detects any modification, insertion, or deletion of data.

- *Nonrepudiation.*   This service can take one or both of two forms: nonrepudiation with proof of origin and nonrepudiation with proof of delivery.

The security services listed earlier can be provided via a number of available security mechanisms. A distinction between the two groups is that security services specify *what* controls are required; the security mechanisms specify *how* the controls are to be implemented. The selection of the appropriate mechanisms is part of architectural considerations. There are three broad types of mechanisms: encryption, access control, and others. Mechanisms can also be classified as preventive, detective, and recovery.

5.3.5 OSI security mechanisms

Security mechanisms include the following:

- *Encipherment.*   Encipherment (also called encryption) is a key security mechanism that can provide confidentiality of either data or traffic flow; often it is used in conjunction with other mechanisms. Cryptography is the discipline involving principles, means, and methods for the mathematical transformation of data in order to hide its information content, prevent alteration, disguise its presence, and/or prevent its unauthorized use.

- *Digital signature.*   This is an encryption method that provides authentication (namely, verification that the individual is who he or she claims to be) and data integrity (namely, the guarantee that data has not been altered or destroyed in an unauthorized fashion). The digital signature is data appended to (or a transformation of) a PDU that allows a recipient to prove the source and integrity of the data.

- *Access control.*   These mechanisms use the authenticated identity of an entity, its capabilities, or its credentials to determine and enforce the access rights of that entity. These mechanisms ensure that only authorized users have access to the protected facilities.

- *Data integrity.*   These mechanisms ensure that the data has not been altered or destroyed. These security mechanisms deal with two aspects of data integrity: the integrity of a single data unit or field and the integrity of a stream of data units or fields. Different mechanisms are used to provide these two types of integrity services. Two specific tools include cryptographic methods (cryptographic checksums) and noncryptographic methods (e.g., error detection codes).

- *Authentication exchange.*   This mechanism provides corroboration that a peer entity is the one claimed. Specific mechanisms include authentication information, such as passwords, supplied by a sending entity and checked by the receiving entity, and cryptographic means.

- *Traffic padding.*   This mechanism provides generation of spurious traffic and filling of PDUs to achieve constant traffic rates or message length.

- *Routing control.*   This is a mechanism for the physical selection of alternate routes which have a level of security consistent with that of the message being transacted.

- *Notarization.* This security mechanism provides the assurance that the properties about the data communicated between two or more entities, such as their integrity, origin, time and destination, are what they are claimed to be.

5.3.6   802.10 LAN security standard

During 1988 the IEEE 802.10 LAN security working group was formed. The group's charter is the development of Standards for Interoperable LAN Security (SILS). IEEE Committee 802.10 is now devoted to specifying data link layer security standards. SILS aims at resolving interoperability problems independent of the encryption algorithms employed by devices on the LAN. This standard aims at complying with the OSI security architecture (ISO 7498-2).

The 802.10 standard includes provisions for authentication, access control, data integrity, and confidentiality. Authentication will prevent stations from reading PDUs destined for a different station; access control limits the use of resources (such as files, servers, gateways, etc.) by unauthorized users; data integrity guarantees that the data is not modified before it reaches the intended user; confidentiality uses encryption to mask the information to users without the appropriate key.

The standard consists of four parts: (1) the model, (2) secure data exchange, (3) key management, and (4) system/security management.

Part A: The Model shows the relation of SILS to OSI; it describes the interfaces and explains which aspects of the interfaces are standardized. Since its formation, the committee has concentrated on development of *Part B: Secure Data Exchange (SDE) Protocol,* to be inserted between the MAC and the LLC sublayers. The working group has also begun development of *Part C: Key Management Protocol* and *Part D: System / Security Management Protocol.* The SDE is a layer 2 protocol that provides services to allow the secure exchange of data at the data link layer. The key management protocol is a layer 7 protocol that provides services for the management of the cryptographic keys used to protect the data at layer 2. The system/security management is a layer 7 set of services that is used to securely manage the security protocols. The SILS model relies on both the OSIBRM security architecture and the OSI management model (ISO 7498-4). The OSI management model describes system management and layer management. SILS aims to use these constructs and relate them back to the IEEE standards environment.

Because WANs are typically implemented using point-to-point links, even when the underlying network is a packet-switched or frame relay facility, the need for extensive security at the data link layer does not arise. In a LAN, connections are intrinsically multipoint-to-multipoint. LANs encompass the concept of subnetworks and routing at the data link layer. Hence, one current school of thought is that increased services at this layer are required vis-à-vis those provided in the OSIBRM security architecture. Some of this rationale is discussed later. Another school of thought is that the international standards from ISO *Network Layer Security Protocol* and the companion *Transport Layer Security Protocol,* provide an adequate apparatus for security. In the discussion below, the IEEE 802.10 view (November 1990, draft 6) is presented. Aspects of the model, LAN security requirements, and the SDE are covered.

Part A: The model. This standard provides the architectural model for security in LANs. The model shows how each of the three LAN security protocols can be used in combination to provide security in a LAN environment. The use of one standard does not mandate the use of either of the other two. This allows implementations to specify compliance to SILS key management, SILS secure data exchange, and SILS system/security management independently. These protocols must support a transparent implementation for devices that currently exist on the network. That is, it must be able to be placed on an existing LAN without causing impairment of communications between already existing devices that do not implement the security protocol.

Figure 5.15 depicts a high-level view of the SILS protocol environment. The user stacks are the existing network communication proto-

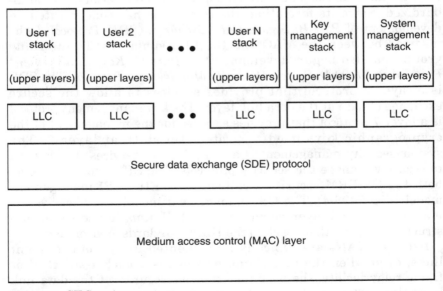

Figure 5.15 SILS environment.

cols before SILS is implemented. These stacks request security services from SDE. In turn, the SDE protocol relies on the key management and system management stacks for support information (e.g., data encrypting keys) in order to provide the security services for the user stack. The user stack may be any protocol stack that would normally reside directly above the MAC layer, including TCP/IP systems. System management and key management reside at layer 7, and for the purpose of the model are specific to SILS. Figure 5.16 depicts a more complete view of the SILS protocol environment; in this figure the cross-hashed boxes represent part of the architecture that must be implemented according to local policy (i.e., are not part of SILS). Arrows indicate communication between processes.

Part B: Secure data exchange. In the course of the development of the SDE standard, the LAN security threats need to be identified. Subsequently, security services necessary to counter these threats can be identified and mechanisms for providing the services emerge.

This standard also defines the layers within the OSIBRM where it is appropriate to apply various security services. In ISO 7498-2, data confidentiality is the only security service indicated for the data link layer. The IEEE view has been to include the services of authentication, access control, and data integrity at the data link layer for the LAN context. ISO 7498-2 took a point-to-point paradigm as the point of departure; LANs are intrinsically multipoint-to-multipoint, necessitating a downward migration of the security functionality. Table 5.9 summarizes the OSIBRM security architecture services and the LAN extensions. The discussion below follows the rationale expounded in the SDE standard.

Usually one needs to protect information both at the application layer and any layers at which subnetworks and routing are implemented. In a LAN, an entire subnetwork can be connected through a MAC-level bridge; this subnetwork can be local or remote. While LANs are similar to packet-switched networks at the data link layer, they also exhibit some of the attributes typically found at the network layer of a packet-switched network, including subnetwork and routing functions. The manner in which data is transmitted and received, the nature of LAN address space, and the new geographic dispersion of interconnected LANs necessitate that access control, authentication, and data integrity be added to data confidentiality (i.e, encryption) typically recommended at the data link layer of a network.

In a LAN, data are transmitted on a medium that is shared by every attached device. Every data unit is transmitted to every other station on the LAN and the source of a given data unit is difficult to authenticate. This presents two security threats: unauthorized

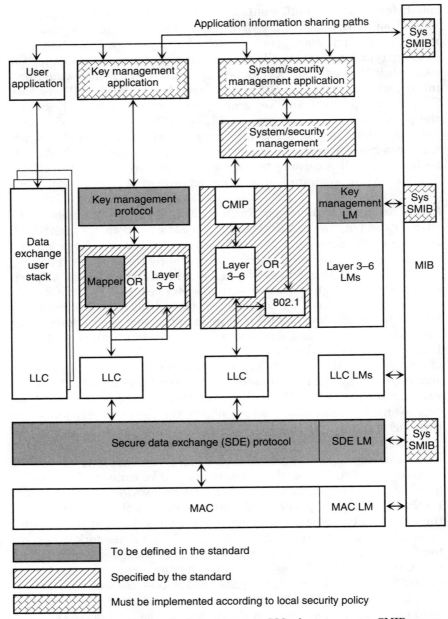

Figure 5.16   IEEE security standards environment. LM = layer manager; SMIB = security MIB.

TABLE 5.9 Security Services at the OSI Layers

	OSIBRM security	LAN extensions
Application	AC, Au, DC, DI, Nr	AC, Au, DC, DI, Nr
Presentation	DC	DC
Session	None	None
Transport	AC, Au, DC, DI	AC, Au, DC, DI
Network	AC, Au, DC, DI	AC, Au, DC, DI
Data link	DC	AC, Au, DC, DI
Physical	DC	DC

AC = access control; Au = authentication; DC = data confidentiality; DI = data integrity; Nr = nonrepudiation.

Note: In an actual implementation, a security service can be included in any layer at which it is listed in ISO 7498-2. A service may appear in one layer, more than one layer, or not at all. The security architecture only specifies where the service can appear, not where the service must appear.

resource use and masquerade. Any station on the LAN can transmit to any other station on the LAN; there are no explicit controls on access to a resource connected to the LAN. Any station can claim to be another station. In terms of reception, there are two threats: unauthorized disclosure and data modification. A station could receive data for which it is not authorized; this station could change the data before it is retransmitted to the intended destination. This is particularly a problem in LANs employing a ring topology such as the 802.5 MAC or FDDI.

Unauthorized resource use and masquerade threats can also come about from the addressing mechanism of the LAN. Each station interface is permanently assigned a specific address. Since any station interface can be attached to any other station interface, a station cannot determine whether the source address of a data unit is legitimate. Hence, any station on the LAN can transmit to any other station. Also, since it is difficult to identify the source of a given data transmission, one station can claim to be another station.

LANs, particularly campus LANs (FDDI) and highly interconnected LAN communities are vulnerable to eavesdropping or tapping. In addition, some LAN media radiate such a strong signal that physical tapping is not even required. This can result into unauthorized disclosure and possibly data modification.

Need for protection. A security service within a layer protects only the SDU, that is the data portion of that layer's PDU. If data integrity is provided at a higher layer, the header information from that layer and all lower layers is left unprotected. In addition, PDUs that originate and terminate within the data link layer (for example, for network management purposes) are also unprotected by the security services provided at higher layers. This is the backdrop for the development, on the part of the IEEE 802.10 committee, of the secure data exchange standard for the data link layer of a LAN, particularly since implementations of security at upper layers are developing too slowly to address some users' needs. Emerging LAN security devices based on the SDE can address these needs until upper-layer security is available.

LAN security service requirements. ISO 7498-2 defines *connectionless data integrity* as "the property that the data in a single connectionless PDU has not been altered or destroyed in an unauthorized manner." This service defeats undetected modification of the protected data (i.e., the SDU). In view of the unauthorized disclosure and modification threats described above, this service is needed for high-security LANs and sensitive applications.

Data origin authentication precludes a station from masquerading as another to illegally access servers on the LAN. Authentication assures a receiving station that the SDU portion of a PDU came from the station whose source address is contained in the PDU header. Data integrity is also necessary as a supportive service: assurance of authenticity of the source without assurance of the integrity of the source address is of no value. When authentication is provided at a higher layer, the header data from that higher layer and below is left unprotected. If an unauthorized station masqueraded as an authorized station and replayed the data contained in a valid PDU, it could result in delivery of data to a station not authorized to receive that data.

Access control. Access control prevents unauthorized use of resources. This is a generalization of what people think of as passwords. Access control provides assurance that access to a resource is granted only to authorized users for authorized purposes. Access control can be applied at either the source or at the destination. The drawback of applying access control at a PDU's destination is that the data has been transmitted to all LAN stations before the transaction is authorized, causing unauthorized depletion of network bandwidth and receiver processing resources, which can result in service denial. Also, access control applied at the destination allows transmissions of data to stations not authorized, whether these stations are directly on the LAN or able to tap in. For example, this service can limit access to a given file server or prohibit access to a communication gateway.

Data confidentiality. Data confidentiality prevents unauthorized disclosure of the protected data. This assures the sending station that the protected portion of a PDU will be available only to the intended recipient. This service is already as appropriate for layer 2 as specified by the OSI service architecture.

Part C: Key management. Key management refers to the storage and proper distribution of keys for secure data transmission. In the proposed SILS context, this application uses the services supported by the SILS key management protocol. Key management is a layer 7 function. It provides keys that are used by the SDE sublayer to provide its services. An authentication mechanism must be included in the key management procedure to ensure appropriate distribution of keys. Owing to government restrictions on export security equipment, the SILS separates the encryption algorithms from the key management algorithms.

Part D: System management. System management monitors and controls protocols across the network at each of the seven layers. The monitoring and control functions are accomplished by communication of information from remote systems using network management protocols. These communications may utilize a protocol such as IEEE 802.10, or they may use the protocol provided at OSI layer 7 (CMIP). While the IEEE 802.1 protocol is sometimes modeled at layer 7, it resides directly above the logical link control layer and can only manage it and protocols below it. This is the reason for the need to develop a more comprehensive LAN standard. Entities responsible for layer 7 communication between end systems are called system management application entities (SMAEs). While system management is concerned with the entire network, layer managers (LMs) act directly at a single layer. For each peer-to-peer protocol at a given layer, there is a layer manager associated with that protocol. The LM may communicate with other LMs at its layer (e.g., IEEE 802.1) to manage the layer. The LM manages the objects used by the protocol at the same layer. Since the LM only has knowledge of events at its layer, it is not capable of making adjustments based upon the needs of other layers. Application processes running at layer 7 can use the SMAEs to exchange information about each protocol layer in different end systems. These application processes then use this information to direct the LMs. The communication of the LMs and the SMAEs is a local matter and is defined internally by an end system.

Examples of objects for communications protocol are window sizes, timers, and buffer sizes. For security protocols, objects are designed to support the security mechanisms. For instance, if encryption is used, a managed object might be the record of the number of significant events such as intrusion attempts. Objects are stored in the MIB. To further depict the protection and separation of security-related objects from other management objects, the concept of a security MIB (SMIB) is introduced in the LAN standard. The structure of the SMIB or the MIB is a local issue; however, the structure of the objects is standardized and is defined in the SMI.

Status of standards. At the time of publication, Part A was being refined to further clarify the infrastructure and the relationships of SILS to the OSI security architecture and OSI network management. Part B is now a standard but it is purely a U.S. standard, not for international usage. However, a new work item was under consideration to extend the OSI security architecture to include the concerns raised in the development of the SDE, discussed above. Part C is targeted for completion in 1993. Part D was being balloted. In the meantime, in the international arena, work has progressed in earnest on the ISO *Network Layer Security Protocol* (draft international standard at press time) and the companion *ISO Transport Layer Security Protocol* (a 1992 international standard).

References

1. IEEE STD. 802.1D-1990, Media Access Control Bridges, IEEE, New York, March 8, 1991.
2. G. H. Clapp, "LAN Interconnection across SMDS," *IEEE Network Magazine,* September 1991, pp. 25 ff.
3. P. Cope, "New LAN Bridges Adapt to Changing User Needs," *Network World,* May 11, 1992, pp. 1 ff.
4. E. Ericson, L. Ericson, and D. Minoli, *Expert Systems Applications to Integrated Network Management,* Artech House, Norwood, Mass., 1989.
5. A. Wittman, "Examining the Ins and Outs of SNMP," *Network Computing,* December 1992, pp. 130 ff.
6. T. A. Cox et al., "SNMP Agent Support for SMDS," *IEEE Network Magazine,* September 1991, pp. 33 ff.
7. J. Case, M. Fedor, M. Schoffstall, and J. Davin, "A Simple Network Management Protocol," RFC 1157, Internet, May 1990.
8. S. Fisher, "Developers Reveal Heartier SNMP," *Communications Week,* June 8, 1992, pp. 1 ff.
9. D. Minoli, "Building the New OSI Security Architecture," *Network Computing,* June 1992, pp. 136 ff.
10. D. Minoli, "What Are the Standards for Interoperable LAN Security?" *Network Computing,* June 1992, pp. 148 ff.

Network Management Issues*

The last few chapters concentrated on first-generation LAN protocols and standards issues. The next few chapters similarly address second- and third-generation LAN standards. As covered elsewhere in this book, studies have shown that the cost of managing a real network can be as high as 40 percent of the total network expenditures. These studies also show that the recurring cost for the communication services utilized, particularly to support WAN-connectivity, is typically around 30 percent of the total network expenditures (the balance being in the amortized expense for communication equipment). Hence, this chapter and the next one take the opportunity to provide a *practical* assessment of these two topics. Naturally, only a selected assessment is provided, since each topic can easily result in a multivolume treatise.

The discussion below follows the Open System Interconnection network management nomenclature shown in Table 6.1.

6.1 Fault Management

Fault management deals with problems associated with physical failures. Failures cause customer inconvenience as well as reduction in productivity and monetary loss for an enterprise. The failure can originate from within the network's immediate environment, or outside of it. Examples of problems include:

---

*This chapter was contributed by J. A. Ladyka, Jr., president, Software Engineering Techniques.

TABLE 6.1   Open System Interconnection Definition of Network Management

Fault management	The process of detecting, diagnosing, bypassing, repairing, and reporting on network equipment and service failures.
Accounting management	The process of establishing charges for the use of communications resources, and costs to be identified for the use of those resources.
Configuration management	The process of maintaining an accurate inventory of hardware, software, and circuits and the ability to change that inventory in a smooth and reliable manner in response to changing service requirements.
Performance management	The process of evaluating the behavior and effectiveness of resources and related communications activities; the utilization of network resources and their ability to meet user service level objectives.
Security management	The process of controlling access to both the network and the network management systems. In some cases it may also protect information transported by the network from disclosure or modification.

- Network interface card fails to operate
- Transceiver fails to operate
- Poor cable plant
- Poor connection at the cable interface with a device
- An internal hardware failure at the file server
- An internal hardware failure at one or more workstations on the network
- Loss of electrical power

This list is by no means exhaustive. The accelerated deployment of LAN-based devices in support of computer-aided workflows throughout an enterprise (e.g., office automation, CAD/CAM/CIM, etc.) has added impetus to the establishment of practices and procedures to minimize outages and disruptions because of a fault.

6.1.1   Cable-related faults

A majority of LAN problems related to communication center around cables. While cabling is about as glamorous as plumbing, it is the very backbone of the enterprise's ability to communicate. Therefore, as part of the planning and implementation of a network, the physical media used in interconnecting network nodes should receive special attention.

A poorly planned and installed cable plant will carry with it an associated high cost during the network's life cycle. This becomes apparent when additions, relocations, and changes are performed on the network having many different media such as twisted-pair, optical fiber, and various types of coaxial cables. Troubleshooting could take hours if the cable run was not documented properly. Given today's high labor rates, it is cheaper to install a new cable than trace an existing cable to see if it is defective.

By having a well-planned and carefully installed cabling system, using consistent media sets throughout an enterprise, problems associated with adding or relocating network nodes are reduced. There are many commercially available cable testers to help map network wiring, spot cable faults, and support the needs of network administrators, LAN cable installers, and field service technicians.

6.1.2 Server-related faults

Although performance is an important factor when dealing with file servers, reliability is equally crucial. With low network or system reliability, mission-critical applications such as on-line reservation systems, industrial process control, and banking transactions would be too costly to maintain. This prompted the need for fault-tolerant systems, which have their roots back in the 1950s and 1960s. Initially, one-of-a-kind custom systems were built for such functions as telephone switching and air traffic control. In 1976, Tandem Computers introduced its first fault-tolerant system, known as Non-Stop, making these types of systems commercially available. Fault-tolerant techniques are now employed with ever-increasing frequency to prevent data corruption or loss. The events that cause data to be lost or corrupted include disk crashes, disk controller failure, electromagnetic interference, and power failures.

Some file servers may have one or more of the following forms of fault tolerance built in: (1) redundancy, (2) failure detection, (3) diagnosis, (4) repair, and (5) software recovery.

To support redundancy, a server may incorporate disk mirroring and disk duplexing. Disk mirroring uses one disk controller for writing to and reading from two hard drives. If one hard drive fails, the other can be used, since it is an exact replica. However, if the disk controller fails, both disks would be corrupted. Disk duplexing alleviates this situation by using two disk controllers and two hard disks. To carry duplexing one step further, a technique called *shadowing* is used. Shadowing entails having a second server attached to the LAN receiving all the data of the first server. Should the first server fail for any reason, the second server takes over. Although this represents an expensive form of fault tolerance, it can pay for itself in outage-sensitive situations, where the loss of a server for more than a few minutes can be devastating. One such situation might be the trading department of a financial brokerage firm.

Novell's NetWare NOS implements a form of fault tolerance known as transaction tracking, whereby operations on the network are divided into transactions. Every operation within a transaction must be completed; otherwise none of them is completed. If the server is interrupted for any reason during a transaction, transaction tracking takes over and resets the server to where it was prior to the interruption. In environments where databases are being updated, this helps to ensure data integrity.

6.1.3 Power-related faults

LANs are often at the mercy of unreliable electrical power. The effects of this are data loss, equipment overheating, hardware damage, and downtime. The quality of the electrical power that is generated at the utility plant is corrupted by external influences such as lightning or grid switching. As a result, the transmission of power is interrupted. Furthermore, internal power contamination can occur within the building itself. For example, when office equipment, electrical motors, and kitchen appliances are turned on, the starting inrush requirements may temporarily overload the utility line and create transient voltages, surges, and sags.

These power disturbances may occur frequently, as shown by two classical independent studies. (The situation has not changed appreciably since the time these studies were done.) It turns out that disturbances occur, on average, twice a week for most commercial sites and that a large proportion of these are generated within the building. In 1969, Allen and Segall of IBM Corporation studied power disturbances related to three-phase 208- and 240-volt power sources from the point of termination of the electrical utility service at the building service entrance. The second study conducted by Goldstein and Speranza of Bell Laboratories for the period covering 1977 to 1979 focused on disturbances coming across 120-volt wall receptacles. This study included power disturbances generated from inside and outside the building. From these two studies, the power disturbances were identified as deviations from steady-state voltage values. ANSI documentation states that steady-state voltages for 120-volt service is defined as a continuous source having a voltage range from 108 to 125 VAC for 120 VAC. The disturbances identified from the studies are:

1. *Failures.* Having a zero voltage condition (also known as outage).

2. *Noise.* Oscillations having a frequency range from 400 Hz to 5 kHz and amplitudes ranging from 15 to 100 percent of nominal voltage. This decaying process occurs within a cycle of line voltage.

3. *Sags.* Decreases in line voltage below 80 percent of the nominal

voltage and typically lasting less than 0.25 s. This type of distur-
bance represents 87 percent of total power line disturbances which
result in data being lost.

4. *Spikes.* Very high voltage, short-term disturbances superimposed
on the line voltage waveform. The duration can be from 10 to 500 μs.

5. *Surges.* Increases in power line voltage lasting longer than 500
μs and being 110 percent above nominal voltage. This type of dis-
turbance causes hardware damage.

To help eliminate the problem associated with these disturbances
and improve the reliability, security, and uptime of the LAN, a device
called an uninterruptible power supply (UPS) is used. The UPS is
plugged into the wall outlet and the equipment needing protection is
plugged into the UPS unit. The equipment could include servers,
bridges, routers, modems, and workstations—anything that is
involved in a mission-critical application. The primary purpose of an
uninterruptible power supply is to supply power during a power failure. Also built into these units are power line filtering and power line
conditioning features (these vary among manufacturers).

UPS units can be classified as being one of two types: *standby* and
online. Standby UPS units do not provide truly uninterrupted power.
Built inside these units are monitoring circuits that detect power failures. However, a small amount of time is required to detect this outage and activate the switching relay to provide standby power. This
means the equipment plugged into the UPS unit could be without
power for 2 to 10 ms. This method only works if the equipment power
supply has enough energy to handle this momentary interruption in
power.

Online UPS units offer the greatest level of protection for equipment. Not only is there no switching time during power failures, they
also function as power line conditioners when utility power is available. The conditioning takes place because the equipment is entirely
isolated from the utility line. The power it does get is generated by
the UPS unit. Online UPS units cost substantially more than the
standby kind. The reason for this is that the inverter must provide
the current demanded at the maximum rated load for an indefinite
period of time. This differs with standby UPS units, which only
require that current go through the inverter during a power failure.

A UPS unit can keep a LAN server running through a short-term
power failure; however, if the power failure becomes protracted it will
provide enough time for an orderly shutdown. This process of orderly
shutdown has prompted the need for an add-on product known as a
UPS monitoring system. The principles of operation involve a UPS
unit providing signals that can be read through an input port on the

processor in question (host computer, server, router, etc.). When a power failure occurs, the processor is alerted to this fact by the UPS unit. The processor then initiates appropriate actions to prepare for eventual battery exhaustion. Typically, these actions include sending warning messages to active users, writing memory caches to disk, and handling the operating system. With a UPS monitoring system installed on a network server, the following potentially dangerous situations can be avoided:

- A user not properly trained in performing a required shutdown is involved in the task.
- The system is running unattended and requires attention.

6.2 Accounting Management

Most networks are comprised of a finite, but often large, number of resources. LAN resource accounting may or may not be required, depending on the situation and the organization. (In contrast, it is usually required for WAN situations.) The network administrator needs to determine if, when, and how to implement resource accounting. Generally, in a LAN environment where there are more users than workstations, resource accounting can help make the users aware of those limited LAN resources and thus encourage efficient use of them. Here are two cases where the choice of account management is clear.

Company A has a LAN with a single server connecting many departments. Each department is considered an accounting entity. For example, if the engineering department buys a storage subsystem and makes space on it available to the drafting department, it can use the resource accounting features provided by the network operating system (such as Novell's NetWare) to track the drafting department's use of the server. The resulting data helps both the engineering department and drafting department with their respective budgets. Therefore, since the engineering department incurred the expense of the server, the engineering department can use the resource accounting data to seek compensation from the drafting department.

On a university campus, students may rent time on a LAN-based system. This entails renting not only a workstation, but an entire system that includes software, hardware, and physical facilities. Here resource accounting provides a mechanism for fairness among students needing access to limited computing resources.

There are times when the choice of account management may not be as obvious because there are no clear indicators, there may be conflicting indicators, or LAN resources may be used unevenly. For example, several key resources may be used constantly, while the remaining resources get used less frequently. To compensate for this, the network administration may set up formal accounting procedures to ensure the availability of needed resources.

6.2.1 Different ways of charging for services

The charging method selected depends on how well the network administrator knows the network environment under which the accounting system will be implemented. The optimal accounting system often comes about by experimentation. Also, how users react to resource accounting will be part of the final decision as to what charging method to use. Below are examples of different charging methods.

1. *Blocks read.*   Charges users every time a workstation reads a block of data from the server. (Blocks are storage structures that represent a pattern or arrangement of data in some physical storage device or medium.) This method of charging is effective if the network is being used as an information service, such as a commercial database. However, using this method may penalize users who need to download large applications from the server.

2. *Blocks written.*   Charges users every time a workstation writes data to the server. This method of charging is effective when dealing with applications, such as spreadsheets, word processors, and personal planners. There are, however, applications such as database management packages that continually write to the disk, thus penalizing the user. Users may also choose not to save their work as frequently.

3. *Connect time.*   Charges by the amount of time a user is logged on to the network. Users may consider this as an implied encouragement to work quickly. However, if accuracy suffers and users need to slow down in order to be more careful, then this type of charge is not appropriate.

4. *Disk storage.*   Charges users by the amount of disk space a user utilizes. This method is suited to networks that have limited disk storage. It helps prevent indiscriminate dumping of files onto the server by users. This charging method helps to promote efficient storage practices but tends to decrease the server's response time.

5. *Service requests.*   Charges users every time a request for service is made to the server. These requests could entail logging in to the server, reading files on the server, sending e-mail messages, and so on. This method of charging is appropriate for networks having an overworked server. Compared with other types of charges that

are specific in nature, service requests charge the user for any type of service. The result of including all service requests into one category is that users tend to economize their requests of the server.

6.2.2 Setting up account balances and restrictions

This section assumes that the network under discussion has Novell's NetWare installed. The discussion is provided to show specific capabilities of typical first-generation LANs. As a network administrator knowing about the SYSCON utility that comes with NetWare, you have the ability to set up account balances and restrictions for the users on the server. The account balance and restrictions option in the SYSCON utility includes an account expiration date, account balance information, password restrictions, and the number of connections allowed for a given user.

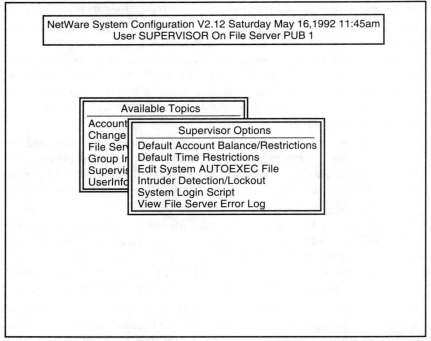

Figure 6.1 Select Supervisor Options from the Available Topics window by highlighting the option and pressing <Enter>.

Restrictions. As part of a network administrator's responsibility, restrictions are applied to users to support the security measures built into the system. To establish these restrictions the network administrator logs on to the server as supervisor, types SYSCON, and presses <Enter> at the prompt. A window titled "Available Topics" will appear, showing six options, one of which is "Supervisor Options" (see Fig. 6.1).

The first option listed on the Supervisor Options window is Default Account Balance and Restriction. Selecting this option brings up another window (see Fig. 6.2).

The changes made to the Default Account Balance/Restriction window only affect users you add onto the server afterward. Previous users retain their prior settings. If the network administrator sets up restrictions before putting any users on the server, those restrictions will then apply to every user eventually put on the system.

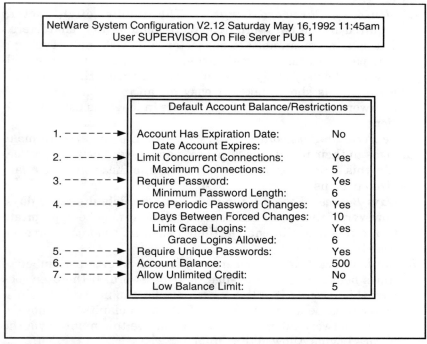

Figure 6.2 The Default Account Balance/Restrictions window lists seven line items to which the network administrator must answer yes or no. These entries are followed by a field where changes can be applied to NetWare's defaults.

Default Account Balance/Restrictions window. These are the options available in this menu.

1. *Account Has Expiration Date.* Allows the network administrator to set a date for a particular account that is to be disabled. If "Yes" is entered here, the next line item, "Date Account Expires:" becomes available for a date to be entered. When that date arrives, the user will not be able to log on to the server. For situations where there are large groups of users coming and going regularly, such as on a college campus, this option facilitates the management of who is allowed on the network.
2. *Limit Concurrent Connections.* Entering "No" to this means users can access the server through any number of workstations at the same time. If "Yes" had been entered along with "Maximum Connections" set to 1, users would no longer be allowed to have more than one active connection to the server at any given time.
3. *Require Password.* The next seven line items deal with passwords and logging on. For security purposes the network administrator should require everyone to have a password; therefore, enter "Yes."
   a. *Minimum Password Length.* The number of characters required in the password. (NetWare defaults to 5 characters.) The maximum number of characters allowed is 20. Typing a long password can be time-consuming and frustrating, especially when the user mistypes it two or three times. Having a password that is short makes it easy for an intruder to guess. A password made up of 6 to 8 characters in length is usually sufficient.
4. *Force Periodic Password Changes.* Users are required to make changes to their passwords every so many days. The primary reason for this action is to prevent intrusion into the system by unauthorized persons.
   a. *Days Between Forced Changes.* NetWare defaults to 40 days; however, if the server has sensitive data, it may be appropriate to require users to change their passwords monthly or even weekly.
   b. *Limit Grace Logins.* Users can log on a certain number of times before having to change their password. If the user goes past this number it will be necessary for the network administrator to issue a new password. NetWare defaults to 5; however, the network administration can change this number via the "Grace Logins Allowed" line item.
5. *Require Unique Passwords.* Can be used to require users to specify passwords that have never been used before every time they change passwords. If "No" is the response to this item, users will

not be required to use unique passwords. Note, using the same passwords over and over makes the system less secure.

6. *Account Balance.* For this line item and the next two to be used, NetWare's accounting services must have already been installed on the server. It is here that the network administrator can set limits on how much service a user can have access to. As shown in Fig. 6.2, the account balance is set to 500 (line item 6); line item 7 does not allow for unlimited credit; and the last item prohibits users access to services if the account Low Balance Limit drops below 5.

6.2.3 Account Restrictions for User window

There will be times when the needs of a particular user will require special treatment. Changes such as these are made to optimize that user's specific networking environment and overwrite the default settings. If it becomes apparent that many individual parameters need changing, the LAN administrator should reconsider how the defaults are defined in the Default Account Balance/Restriction window (see Fig. 6.3).

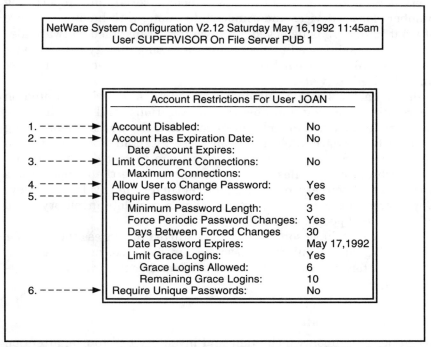

Figure 6.3 When you select the account restrictions option from the User Information window you are presented with the Account Restriction for User window as shown above. This is where the network administrator sets up individual restrictions based on personal needs.

6.2.4 Tying it together

Just as a CPA must design an accounting system that fits the idio-
syncrasies of a particular company, so must a network administrator
design a resource accounting system around the requirements of the
company and the needs of its users. Good planning implies knowing
the hardware and software so that the above criteria can be met without waste of capital and staff time.

6.3 Configuration Management

When LANs first appeared in the late 1970s and early 1980s, they
consisted of a small set of PCs cabled together to form primitive
departmental systems. Since that time, enterprisewide networks
involving thousands of PCs have been deployed. With the associated
complexity that is inherent with such large systems, the network
administrator is often called upon by corporate management to provide reports (cost, security, planning, etc.) on resources deployed on
the network for control purposes. This means every workstation, server, bridge, router, hub, and gateway and the corresponding configuration must be known. This could also include serial and property-tag
numbers for all devices and components. Along with having a picture
of what hardware is deployed, an inventory of installed software also
needs to be maintained to ensure upgrade compatibility between versions, compliance with license agreements, and security against
unauthorized software.

   In the past, collecting this data usually meant crawling around on
floors, disconnecting and opening workstations, and recording the
components attached to and contained within the workstation. Trying
to identify software applications on these PCs and workstations can
be laborious. Some systems may store them in root directories containing hundreds of files, while others may have their applications
buried in obscurely named subdirectories, many levels deep.
Depending on the size of the corporation, a complete inventory could
take a team of people weeks of effort.

   Only recently have software vendors started to address the need for
automating the inventory process and, in doing so, helped network
administrators with their documentation and configuration management requirements.

6.3.1 How to get started

When deciding upon which audit and inventory tool to use, attention
must be given not only to how the software package tracks hardware
and software configurations, but how it handles other features as well.

These features might include programmed alerts, current status, history detail, multiserver environments, and report generation. Each product has its own strengths and weaknesses, which must be considered in conjunction with the specific requirements of the network administrator. The following is a brief description of such features:

- *Alert function.*   Presents simple messages on the console as to changes that have occurred; for example, "Amount of free memory has decreased by 300 KB to 340 KB."

- *Current status function.*   Provides information about a given node on the network and records it in a *current status* database.

- *History function.*   Maintains a complete history of each node. Whenever a change occurs, a snapshot is taken and placed in the history database.

- *Multiserver support function.*   Enables network administrators responsible for networks covering a large number of sites and with many servers to track users' names and the servers on which they logged in.

- *Report generation function.*   Provides a (simple) tool with which meaningful and easy-to-read reports can be generated from the data collected. Usually, no single package generates all the reports a network administrator might want.

6.3.2   Details found in configuration information

For effective network management it is necessary to know exactly how each workstation is set up. The configuration information related to these workstations can be partitioned into two areas: hardware and software.

Hardware.   Attributes of the machine, such as:

- Type of microprocessor (80286, 80386SX, 80386DX, etc.)
- Type of numeric ("math") coprocessor (80287, 80387, etc.)
- Type of bus (ISA, EISA, Micro Channel)
- Total amount of random access memory (RAM)
- Total number of storage devices (hard disk drives, optical drives, etc.)
- Type of mouse (serial or bus)
- Type of video (EGA, VGA, SVGA, etc.)
- Total number of serial and parallel ports
- Type of cards in each slot inside the machine

- Basic input-output system manufacturer and date
- Complementary metal oxide semiconductor (CMOS) errors pertaining to serial and parallel port addresses

Software. Information regarding operating systems, device drivers, and memory management, such as:

- Types and version level of operating systems (Microsoft DOS, IBM DOS, IBM OS/2, etc.)
- Configuration of the operating system
- Types and versions of memory device drivers (expanded memory specification, XMS, HMA)
- Random access memory available and free for use
- Type and version level of LAN driver

The above set of hardware and software configuration elements is tracked by many of the available management packages.

6.3.3   Workstations having local disk drives

For configuration management, inventory purposes, and software license metering, it is important to be able to audit the software on the local disk drive. The ability to list files based on manager-selected configuration options and to count the number of times a given file occurs are desirable features. The ability to track and modify text files on workstations, allowing for customizable audits of the contents in those files, by some specified criteria, is another desirable feature. Files such as CONFIG.SYS may take less than a second to examine, while .EXE and .COM files may take longer. Programmable auditing intervals is another feature provided by many auditing packages to accommodate examining the contents of files.

The local disk drive itself includes information as to the number of heads, sectors, cylinders, and tracks. The drive also has information regarding how it is logically partitioned, such as DOS bootable or nonbootable, OS/2 HPFS (high-performance file system). Therefore, if the contents of the CMOS memory on the workstation become lost, the network administrator should be able to go back to the information captured by the auditing package and make the drive accessible once again.

6.3.4   Inventory information

To facilitate the management of the LAN infrastructure, the configuration management package should be able to track individual workstations as well as the hardware components that make up the workstations. The facts to be included with this are as follows:

- Manufacturer's name
- Manufacturer's model and serial number
- Purchase date of the component
- Name of vendor
- Vendor's phone number (this may be a user-definable field)
- Vendor's contact person (this may be a user-definable field)
- Cost of the component
- Warranty expiration date
- Maintenance contract information

This information could help the network administrator in the budgeting process, providing technical support and upgrading the network when necessary.

6.3.5 List of audit and inventory packages

Table 6.2 supplies an abbreviated list of auditing products (for illustrative purposes only).

6.4 Performance Management

Owing to the rapid penetration of communication-based workflows in the corporate environment, networks have become an integral part of most corporations, whether they are large, medium, or even small enterprises. As the number of users increases, so does their dependence on the LAN. Therefore, the network administrator must keep the network running at a reasonable level of performance and repair it immediately when it is not.

The tools that are used to evaluate network performance are generally known as *network analyzers* and *protocol analyzers*. There is a subtle distinction between the two products. Protocol analyzers are debugging tools originally developed for software engineers with

TABLE 6.2   Partial List of Audit and Inventory Products

Product	Vendor
Network HQ	Magee Enterprises, Norcross, Georgia
LAN Auditor	Horizons Technology, San Diego, California
PC Census	Tally Systems Corp., Hanover, New Hampshire
LAN Automatic Inventory	Brightwork Development, Tinton Falls, New Jersey
Argus/N	Triticom, St. Paul, Minnesota

extensive technical experience. The data consists of statistics about the traffic transacted over a network. Network analyzers provide a less technical network administrator with user-friendly interfaces and interpreted information. This tool provides a proactive mechanism to localize performance problems before they can cause a serious degradation of the network's throughput and response time.

Both the protocol analyzers and network analyzers should be part of a network administrator's toolbox. They can be used to identify (trace) which nodes are sending out which traffic or whether one node cannot communicate. The network administrator can also gain an accurate picture of the traffic on the LAN. For example, does the traffic peak at certain times of the day? Has someone put a node on the network without prior authorization? Does the network need to be segmented and then interconnected with a bridge?

These are just a few of the questions a protocol analyzer can help answer. However, it takes experience on the part of the network administrator to know what a malfunctioning transceiver looks like on a protocol analyzer and also be sophisticated enough to set up the filters and scripts needed so that only the necessary information (rather than hexadecimal dumps of packet decodes) is provided. A network analyzer, by contrast, is easier to use and provides many more clues as to what the problem might be. For the most part, network analyzers are used for troubleshooting and capacity planning.

6.4.1 Protocol analyzers

The previous chapters provided extensive coverage of LAN protocols. This section takes a more intuitive view of protocols, not as much from a developer's perspective, but from a practitioner's perspective.

Protocols are formal rules governing communications between entities (workstations, servers, bridges, routers, etc.). In order to communicate, certain steps must proceed in the right sequence. A computer workstation on a LAN must gain access to the physical medium. This is managed by the network interface card installed in the workstation, PC, or server. Ethernet networks utilize contention access. Requests for use of the network resources (bandwidth) are generated each time information needs to be transmitted. If simultaneous requests are made by two or more workstations on the network, collisions will occur. Each workstation must wait a random amount of time before reattempting transmission. Prior to transmission, the user's data is enveloped with protocol control information, which includes addressing, as discussed in Chaps. 3 and 4. If the data is traveling over a WAN, routing information appropriate to that WAN must be included (typically as network layer PCI). Each LAN protocol has a specified format, as discussed in Chap. 3. For example, the

MAC PDU includes, among other fields, frame synchronization octets and error-checking bits. If any part of the frame is not in a format recognizable by another workstation or server, the two cannot communicate.

As networks grow and become more heterogeneous, it is more difficult to ascertain that compatible protocols are supported by all network entities. Equipment vendors may develop protocols that are slightly different from published standards and from one another. Others may develop proprietary protocols. Having a homogeneous network does not necessarily ensure interoperability. A network interface card may suddenly generate bad frames or not generate frames at all. The only way to isolate a network failure of this type is to examine the bits that make up the frame. The protocol analyzer allows for sophisticated analysis of frames traveling over the network, but it cannot warn of impending network failure as can be done with network analyzers. Therefore protocol analyzers are used primarily as a diagnostic tool.

Originally, these devices were used by software and hardware developers to ensure that their products conformed to a specification. However, the only way to do this is to examine each frame bit by bit to locate the problems. Traditionally, protocol analyzers have not been considered proactive tools, as is the case with network analyzers. They tend to be kept in a closet by the network administrator until the network has a problem. Today's protocol analyzers do include features that allow the network administrator to take a more proactive approach and thus be made aware of a problem at an early stage. By using such tools on a regular basis, the network administrator can become familiar enough with the performance characteristics of the network to be able to forecast if something is going to cause the network to fail.

Filters and triggers can be used to capture data on the basis of its characteristics. The terms *filter* and *trigger* refer to the type of screening to be performed on network traffic. A trigger can be used to start or stop the process of collecting data. For example, a trigger can be set to stop collecting data as soon as an error condition has been detected. Then the network administrator can go over and examine the data before the error condition occurred and make note of the pattern that caused the failure. A filter is similar to a trigger in that it collects specific data. It continuously screens only for specific data needed, such as parts of frames or frames coming from a certain source or going to a certain destination. If a particular set of protocols needs to be examined, a filter can be made to capture only that set and ignore the rest. Through the use of these features and the ability to detect and record network conditions, the network administrator can collect information on network performance. From this historical baseline, a better understanding of the traffic patterns can be achieved by identifying possible deviations.

The implementation of filtering can be achieved by hardware or software; hardware filtering is faster, but software filtering is more flexible. The combination of both hardware and software filtering provides the best arrangement. The hardware filter may have a limit on the number of frames that can be captured; in this case the capturing may be handled by the software. Without losing frames, the software can be used to fine-tune the filtering process and provide only relevant information to the network administrator.

A protocol analyzer's price ranges between $10,000 and $25,000. Some provide the ability to export information to other applications such as Lotus 1-2-3, FoxPro, and Paradox, so data can be manipulated. Other analyzers have emulation capabilities. Here the network administrator can generate traffic that looks like it is coming from a particular workstation. Having this ability becomes important as the network grows in complexity, since it allows testing for problems that have not been anticipated. Because these devices are relatively expensive and may be used infrequently, renting one may help reduce the cost, particularly for small enterprises. Depending on how critical the network is to the business, the network administrator can estimate the financial cost of the downtime and use that to justify the purchase of a protocol analyzer.

6.4.2 Network analyzers

In today's environment of interconnected LANs, distributed computing, and heterogeneous networks, network analyzers are becoming an integral part of the repertoire of network management tools. A network analyzer can be used to spot address conflicts, congestion at bridges and routers, and incorrect software configurations which might affect communications. An analyzer can also detect bad frames and FCS errors. With the advent of multiprotocol enterprise networks and interconnection of LANs over WANs, network analyzers have evolved to become hybrid devices combining network monitoring, historical trend reporting, and protocol analysis.

A network analyzer can come as one of three types: software, kit, and stand-alone. The software version uses the network interface card within the workstation to monitor the network. The kit utilizes a special network interface card and software. Both the software and kit type analyzers use the processing and storage capabilities of the workstation they are installed in. Stand-alone types come as portable computers.

To help make the use of an network analyzer easier, vendors have added intelligence to their products, such as improved user interfaces and simplified presentation of information. Expert systems have been included in some products to automate the interpretation process.

These devices support the type of distributed analysis now required by network administrators for managing internetworked LANs. Distributed analysis allows for increased levels of integrated management. For example, analysis capabilities can be integrated into smart hubs located on various floors and buildings, and the information can be collected for decision making through SNMP, as discussed in Chap. 5.

A smart hub can determine the source of data through port-to-address mapping. This feature provides the network administrator with the ability to selectively decide which port he or she is going to examine. By placing remote network analysis units on other network segments, a network administrator at one location can look at frames captured in another location. If this setup is done correctly, remote frame capture will send frames back to the central station prior to decoding; this reduces the traffic on the network. Having the capability to do analysis in remote units provides the network administrator with the ability to identify the type of problem occurring on a particular segment and come up with the right solution for the problem. This becomes important if a remote segment is in a distant town or towns. Through distributed monitoring, a total view of the trends can be seen; for example, the administrator can study how performance changes in one area can affect other parts of the network. Management systems that support SNMP remote probe devices allow the network administrator to:

- Search for irregular behavior without polling all nodes and report any problems back to the network management device
- Analyze the actual packet and frame data for troubleshooting network segments

6.4.3 Maintaining performance

The challenge of keeping large, complex networks running smoothly is complicated by the current state of network management. Many network management tools still have limited real intelligence, thereby forcing network administrators to rely on what they know or can figure out by themselves. Therefore, the following actions may be helpful in maintaining performance to an established level:

- Ascertain timely proactive notification of LAN problems, as well as WAN problems in the case of remotely connected LANs

- Retain information pertaining to normal frame flow, number and types of bridges, routers, servers, and network protocols
- Ensure that the network management information accumulated is maintained for historical baselining

6.5 Security Management

Refer to Chap. 5 for issues pertaining to security in a LAN context. Also refer back to Sec. 6.2.2.

7

Connecting
Dispersed
LANs

At first LANs were islands unto themselves. Now corporations are putting in place enterprisewide networks connecting every department of the organization. New work paradigms involving "cooperative work," as well as streamlining and efficiency in order to deal with the global economy, naturally lead to a requirement for high interconnectivity. This chapter discusses important advances seen in the past couple of years in WAN services available to corporate LAN managers to effect such connectivity. Dedicated high-speed digital services and switched services (frame relay and SMDS) are discussed here; ATM and cell relay are discussed in Chap. 9.

A variety of communication options is available to meet specific users' needs. Just a few years ago, low-speed dedicated communication channels were most commonly used. What about interconnection facilities for LANs supporting CAD/CAM, high-resolution graphics, animation, visual simulations, and other engineering applications? Well, there are communication solutions ideally suited to these applications. Below, we survey some of the communications services which can be employed to achieve distributed data processing and LAN connectivity.

7.1 Plethora of Communications Options to Support Interconnection

The computing environment has evolved from stand-alone PCs in the early to mid-1980s, to locally networked PCs on site and campus-based LANs in the late 1980s, to interconnected enterprisewide systems in the early to mid-1990s. Table 7.1 provides a summary of some

TABLE 7.1 Communications Options to Support "Network Computing"

	Nonswitched	Switched
Low-speed	Analog private line; DDS private line; fractional T1 private line; T1 private line; frame relay (permanent virtual circuit)	Dial-up with modem; ISDN; packet-switched network; frame relay (switched virtual circuit)
High-speed	T3 private line; SONET private line; ATM/cell relay service (permanent virtual circuit)	Switched multimegabit data service (SMDS); ATM/cell relay service (switched virtual circuit)

of the key communication solutions which can be employed by the LAN manager to achieve the desired connectivity.[1] Typically these services are employed between local and remote bridges and between routers, as shown in Fig. 7.1.

High-speed digital services can be classified as *dedicated services,* such as fractional T1 (FT1), T1, and T3, and *switched services,* such as switched T1, frame relay service (public or private), and SMDS. The 1990s are seeing the introduction of new communication services geared to the evolving data-intensive applications which businesses are increasingly bringing on-line to address workforce productivity and efficiency. Factors such as increased throughput requirements (from kilobits per second per station to megabits per second per station),

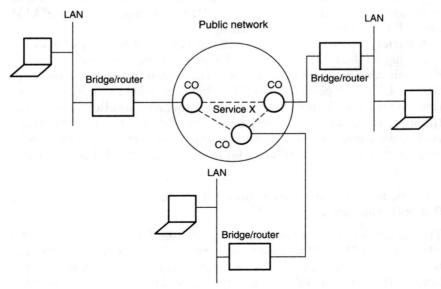

Figure 7.1 Telecommunication service to interconnect dispersed LANs.

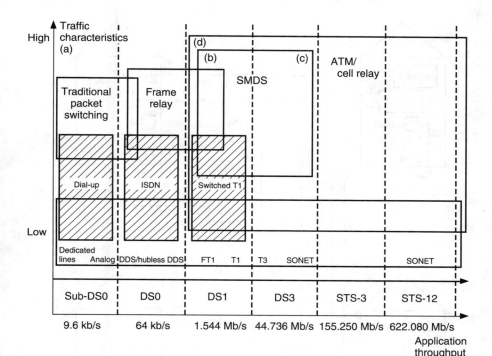

Figure 7.2 Range of available WAN technologies. *a:* Burstiness or bandwidth utilization. *b:* Some are also proposing $n \times 64$ kb/s access. *c:* SMDS will be upgraded to higher speeds when ATM PVC becomes available. *d:* Initially defined at 155 Mb/s, but more recently also defined at DS1 and DS3 access.

increased penetration of terminals in the corporate workforce, and a higher degree of required interconnection across cities, regions, and even continents are making some of the traditional communication services inadequate, marginally adequate, or simply too expensive.

Figure 7.2 depicts the spectrum of available internetworking technologies, parameterized on bandwidth requirements and user's traffic characteristics. For example, frame relay supports bursty traffic at medium speeds. Competing technologies include T1 links, fractional T1, fractional T3, ISDN primary rate service, ISDN H0, ISDN H11, switched DS1, SMDS, and ATM/cell relay. Some of the key bridge-to-bridge and router-to-router services are discussed in the sequel.[1-4]

7.2   Interconnecting LANs over a Wide Area: High-Speed Nonswitched Solutions

Figures 7.3 through 7.6 depict the evolution of corporate networking in general and LAN interconnection in particular.[5] In most cases, connectivity meant the use of dedicated carrier facilities.

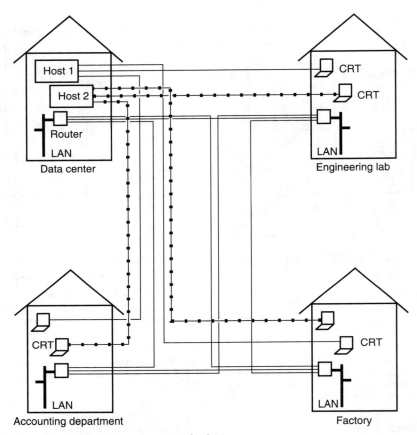

Figure 7.3   Early 1980s unintegrated solutions.

Figure 7.3 depicts the early 1980s unintegrated solution. LAN connectivity, if at all, used its own transmission facilities. Different departmental data applications (for example, a mainframe payroll application and a minicomputer supporting marketing) used separate networks. Not only was this solution expensive because of the duplicate transmission costs, but it was also difficult to manage and grow.

Figure 7.4 shows the advantage of the T1 multiplexers introduced in large numbers in the mid-1980s. The data applications were aggregated over a common backbone network, improving network management, simplifying the topology, and reducing the communications costs, since a few traditional analog lines are enough to justify the cost of a high-speed digital link. One of the shortcomings of this approach, however, is that the LAN traffic usually remained separate, giving rise to two overlay networks. LAN interconnection was

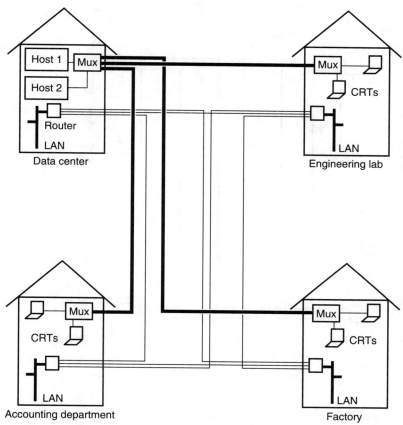

Figure 7.4 Introduction of T1 multiplexers.

still outside the "mainstream" data communication, data processing infrastructure.

Figure 7.5 depicts a more recent situation, where the T1 multiplexer is made to support the LAN interconnection traffic. LANs are brought into the mainstream. Up to this phase the corporate network comprised traditional technologies and interfaces. The cost of T1 links has been going down steadily in inflated and even constant dollars since 1984, with the latest drop in the summer of 1992.

Figure 7.6 depicts the use of public switched service, which is a trend of the 1990s; this approach is discussed later at length.

7.2.1 Analog private line connectivity

This is the classical interconnection method employed by network designers for the past quarter century. It involves a permanently

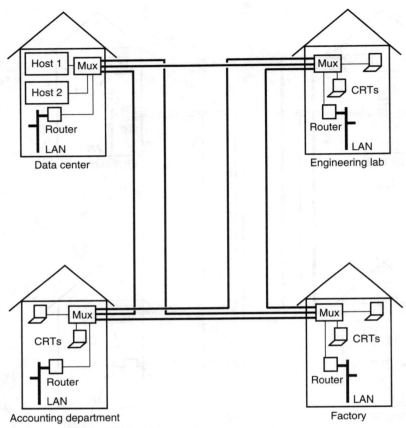

Figure 7.5 LAN over T1 backbone.

installed voice-grade line between two points. These channels are adapted from voice communication to data communication through the use of a modem. The modem transforms data into an analog signal suitable for transmission over the traditional telephone network. While digital backbones are now becoming popular, a nontrivial portion of today's data communications is still carried by voiceband modems over the analog telephone network, particularly for terminal-to-mainframe applications (automatic teller machines, lottery terminals, claims processing, etc.).

This approach, although relatively inexpensive (approximately $80 plus 75 cents a mile per month), only supports a bandwidth in the 9.6 to 19.2 kb/s range.

7.2.2 Fractional T1, T1, T3 and SONET

Dedicated "private line" services provide transparent bandwidth at the specified speed ($n \times 64$ kb/s for FT1, 1.544 Mb/s for T1, and

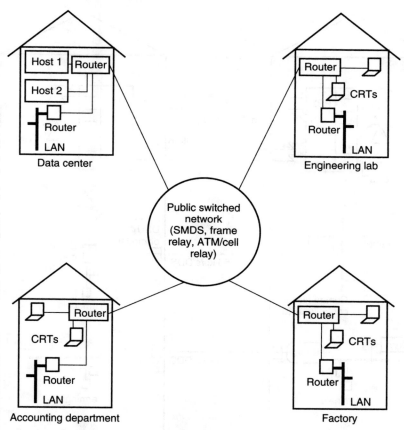

Figure 7.6 Using public switched service. (*Note:* Switched virtual circuit capability needed to make frame relay and cell relay truly switched.)

44.736 Mb/s for T3) and are suitable for point-to-point interconnection of low-burstiness (i.e., steady) traffic. Data submitted at one end is sure to get to the other end, except for unexpected failures of the link and other low-occurrence errors. Figure 7.7 depicts a fairly common approach for connecting LANs located at major customer sites. (This approach is not ideal when there are many dispersed sites.)

A T1 link costs approximately $900 twice (two ends) for access to an interexchange carrier (if any), plus $2500 plus $3.5/mi as an average approximate monthly charge. A T3 line costs about as much as 6 to 10 times a T1 line, while an FT1 line costs approximately the corresponding fraction of a T1 line for the interexchange portion (the access portion remains at $900 per end). As the number of remote sites increases, the interconnection cost becomes higher.

T3 (also known as DS3) facilities are increasingly available in many parts of the United States, although they are still fairly expensive. If

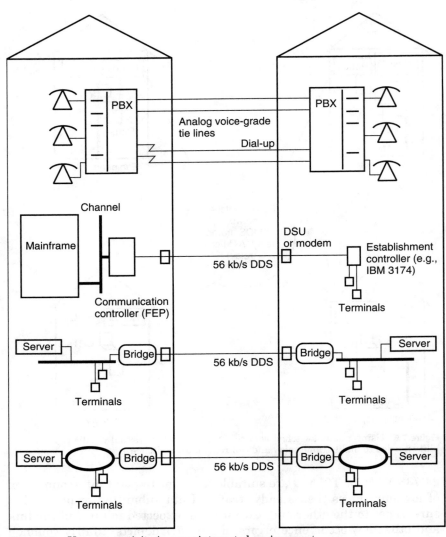

Figure 7.7a User connectivity in an unintegrated environment.

one needs more than 6 to 10 (depending on distance) T1 circuits between two points, then it is cheaper to use a T3 link. In addition, a number of carriers have started to offer a fractional T3 service, which allows the user to specify the desired number multiple of T1s.

A new digital hierarchy, suited to handle high-bandwidth, fiber-based signals, has recently been defined. This standard is known in the United States as *synchronous optical network* (SONET). In the early 1990s, telephone operating companies and interexchange carri-

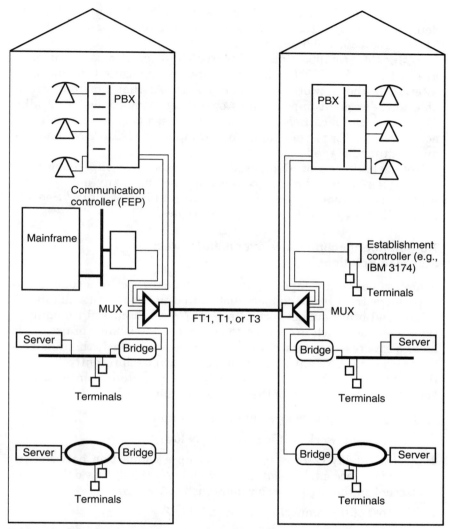

Figure 7.7b User connectivity in an integrated environment.

ers will deploy network equipment meeting the new SONET standards. SONET's hierarchy of rates and formats starts at a 51.840 Mb/s rate. The SONET hierarchy addresses transmission up to 2.5 Gb/s and is extensible, if necessary, to more than 13 Gb/s. The basic building block is a 51.840 Mb/s signal known as a *synchronous transport signal–level 1* (STS-1). SONET defines the basic STS-1 signal and an associated byte interleaved multiplex structure that creates a group of standard rates at n times the STS-1 rate. n takes selected integer values from 1 to 255; currently, the following values are defined: n = 1, 3, 9, 12, 18, 24, 36, and 48 (the corresponding rates are 51.840, 155.250, 466.560 Mb/s, …).

In the 1993 to 1995 time frame, one can expect to see commercial availability of SONET facilities for end-user applications (deployment internal to the network will occur sooner). At that time one will be able to order point-to-point lines operating at multiples of 52 Mb/s. These high-end facilities clearly would be used only where the traffic across the LANs is very high, for example, in CAD/CAM, imaging, and multimedia applications.

Networks based on dedicated services become expensive when many sites have to be interconnected since as many as $n(n - 1)/2$ links may be required, where n is the number of sites.[6] See Fig. 7.8(a). This is why switched solutions are desirable.

7.3   Interconnecting LANs over a Wide Area: Switched Solutions

Seamless network computing requires the existence of a communications infrastructure to connect distributed users, servers, databases, LANs, and other subnetworks. In Sec. 7.2 we discussed communications services which are suited to connect a few islands of users who have high cross-network traffic requirements. In fact, this has been the traditional LAN interconnection approach until recently. This section examines other techniques which are suited to interconnect LANs when one or more of the following conditions hold:

1. There is a relatively large number of islands of users.

2. The cross-network traffic is relatively low.

3. A small delay (2 to 8 s) in setting up a connection between two islands is acceptable (for the low-end solutions; this delay is not present in the high-end solutions such as SMDS).

4. The cost of the connection is important.

The technologies discussed below can be grouped as low speed (e.g., dial-up) and high speed; or as existing (e.g., integrated services digital network) or emerging (e.g., ATM-cell relay); or as low-tech (e.g., dial-up) or high-tech (SMDS). No particular grouping is employed in the discussion which follows.

7.3.1   Dial-up link between LAN communication servers

This approach (also known as circuit switching) involves the use of modems connected to the LAN server (bridge or router) to connect to

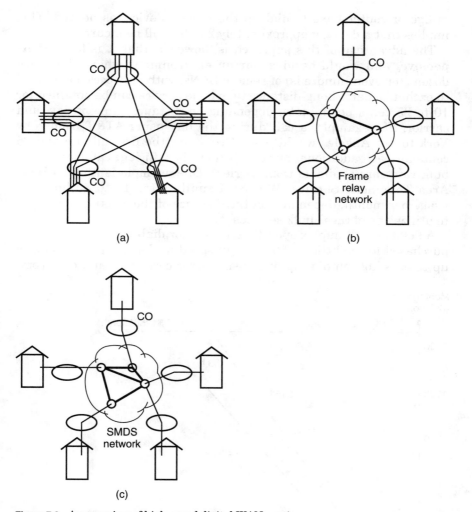

Figure 7.8 An overview of high-speed digital WAN services.

the analog public telephone network. This implies that the communications channel is not dedicated 24 hours per day, but must be brought on line when needed (via a process called *call setup*) and taken down when no longer needed. Traditional modems have operated at speeds up to 19.2 kb/s, but 9.6 kb/s is more common. This implies that the throughput across this type of LAN-to-LAN link is fairly small. Consequently, only a small number of users and short inquiry-and-response transactions can be supported. Since the link between the two servers is not available on a dedicated basis, the bridge or router have to dial up the remote device, as needed. This implies that a delay of approximately 2 to 8 s will be incurred.

The advantage of this approach is, however, that it is fairly inexpensive, and would be ideal for an environment where there are dozens (or even hundreds) of remote LANs, with only occasional need to exchange data. Long-distance telephone service can be obtained for 10 to 25 cents per minute, depending on distance, time of day, and carrier. For example, a call during the day using AT&T from New York to Los Angeles would cost 24 cents for the first minute and 25 cents for each additional minute. If there is sufficient calling volume, bulk rates are available from carriers. One example is AT&T's Wide Area Telephone Service (WATS). Usually after 10 to 30 hours of usage per month (depending on the distance of the calls) it is cheaper to utilize one of these bulk services.

A 9600 b/s full-duplex operation modem for dial-up lines can now be purchased for as little as $600. High-speed modems can now achieve up to 38.4 kb/s on dial-up lines using error correction and data com-

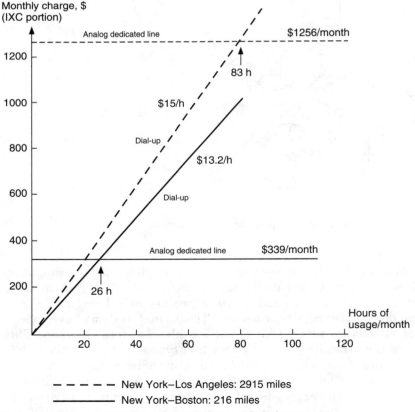

Figure 7.9 Comparing traditional private lines with dial-up service.

pression (some go as high as 57.6 kb/s); these modems now appearing on the market can be purchased for approximately $900.

Figure 7.9 depicts a comparison of the cost of a dial-up link between two LAN bridges in New York and Los Angeles, and New York and Boston, and the cost of a dedicated analog line. As it can be seen, if the monthly usage for the LAN-to-LAN traffic between New York and Boston (Los Angeles) is less than 26 (83) hours, then dial-up is cheaper.

7.3.2 Integrated services digital network dial-up link between LAN communication servers

This approach involves the use of newly introduced switched *digital* facilities between the LAN servers. Integrated Services Digital Network (ISDN) provides end-to-end digital connectivity with access to voice and data services over the same digital transmission and switching facilities. It provides a range of services using a limited set of connection types and multipurpose user-network interface arrangements. ISDN provides three channel types: B-channels, D-channels, and H-channels. The *B-channel* is a 64-kb/s access channel that carries customer information, such as voice calls, circuit-switched data, or packet-switched data. The *D-channel* is an access channel carrying control or signaling information and, optionally, packetized customer information; the D-channel has a capacity of 16 kb/s or 64 kb/s. The *H-channel* is a 384-kb/s, 1.536-Mb/s, or 1.920-Mb/s channel that carries customer information, such as video teleconferencing, high-speed data, high-quality audio or sound program.

ISDN further defines *user-network physical interfaces*. The more well known of these interfaces are:

2B+D   Two switched 64-kb/s channels, plus a 16-kb/s packet-and-signaling channel (144 kb/s total)

23B+D   Twenty-three switched 64-kb/s channels, plus a 64-kb/s packet-and-signaling channel (1.536 Mb/s total)

In addition, some carriers have announced the availability of the following options:

H0 + D   Switched aggregated 384-kb/s links

H10 + D   Switched aggregated 1.544-Mb/s links

As can be seen, ISDN provides considerably more bandwidth on dial-up facilities than standard analog circuits. ISDN is now available in the top 50 markets and is expected to be increasingly available in other areas.

7.3.3 Packet-switching connectivity

Traditional packet switching is a technology which first arose in the mid-1960s. Packet switching affords statistical allocation of bandwidth. Packet switching has been standardized since 1976, using the X.25 set of specifications. Its throughput has traditionally been limited to around 9.6 kb/s and, hence, is not ideally positioned to support today's networked applications, although some users could still employ it, particularly for textual e-mail applications. It is treated here mostly for historical reasons.

In packet switching, information is exchanged as blocks of limited size, or *packets*. At the source, long messages are partitioned into several packets which are transmitted across the network and then reassembled at the destination to reconstitute the original message. Multiple users can share network resources, thereby lowering the costs, although efficient use of transmission resources increases the network complexity. Buffers are needed at each node in a packet network; the storing is typically transient in nature and is typically of the order of tens or hundreds of milliseconds. Packet-switching service can be obtained via a privately owned network or a public packet-switched carrier.

LAN technology is based on a form of packet switching (connectionless mode); hence, interconnection of LANs using packet-switching technology can be efficient, particularly if the packet-switched service is also connectionless mode, such as SMDS. X.25, however, is connection-oriented; refer back to Chap. 4.

Effective throughput of traditional WAN packet-switching service has been at the 9.6-kb/s or 56-kb/s rate. Packet-switching services are also available through ISDN. The X.25 packet standards assume that the transmission medium is error-prone. In order to guarantee an acceptable level of end-to-end reliability, error management is performed at every link using a resource-intensive link protocol such as link access procedure B (LAPB), which is an implementation of high-level data link control (HDLC). This can become time-consuming and consequently affect the end-to-end network latency. Fiber-based transmission facilities now routinely provide a 10^{-9} bit error rate. The amount of bandwidth required by new user applications has increased in the recent past. These new conditions are opening up the door for frame relay, SMDS, and cell relay; these services, which eliminate the performance restrictions of X.25, are discussed below.

Terrestrial private line costs are based on mileage charges and are therefore distance-sensitive. This is true of analog private, DDS, T1, FT1, and T3 lines. These lines are relatively insensitive to the volume of data, once the speed has been selected and fixed. By contrast, packet-based networks are typically priced on a timed-usage basis and are therefore sensitive to data volume; they are, however, insensitive to distance and are only slightly sensitive to the number of sites added. Figure 7.6 suggests what the topology of this solution looks like (although packet switching is not explicitly listed in the figure).

A typical usage charge for a public packet-switched network with access to a packet assembler-disassembler (PAD) at 2.4 kb/s is 25 cents for the first 2 minutes and 0.125 cents for each additional minute. Two local loops will be required (one at each end); these cost on the average $48 per month. For example, an hour of actual usage a day would translate into $7.75 per day, or $170 a month; note that for under 150 miles or so, a dedicated voice-grade line would be cheaper. (If the actual usage was 2 hours per day, then for under 300 miles the voice-grade line would be cheaper.) Packet charges are additional.

CCITT recommendation X.25 is a standard interface protocol between packet-switched user equipment and a packet switch, covering the lower three layers (also called *levels* in this context) of the OSI reference model. The physical layer, or layer 1, is concerned with physical connectivity to the network. The next layer, the link layer, or layer 2, deals with error control and flow between two adjacent points. Layer 3, the network layer, deals with end-to-end networking aspects, including routing. The X.25 recommendation was first adopted in 1976, and was significantly revised in both 1980 and 1984; minor revisions took place in 1988.

The physical level deals with the representation of data bits, timing aspects, and physical contact between terminal [known here as data terminal equipment (DTE)], and the data circuit-terminating equipment (DCE), for example, a modem. These functions are handled by the X.21, X.21bis, X.20, and X.20bis specifications.

The link level protocol used in X.25 is LAPB. The link level provides the functions of link initialization, flow control, and error control. LAPB provides an error-free data channel despite the potential unreliability of the physical medium. At the receiving end of the link, the information is delivered in units (or packets) to the packet level without loss or duplication and in the same sequence of transmission. Specific bit patterns (flags) delimit the data link frame and also provide a means for link level synchronization. An LAPB frame consists of a link level header, an information field (if any) and, to detect transmission errors, a 16-bit frame check sequence. The header consists of an address field and a control field. The header shows whether the frame is a *command* or a *response* and differs by frame type—that is, information frames (I format frames), supervisory frames (S format frames), or unnumbered frames (U format frames).

Every information frame to be transmitted across the interface is sequentially numbered in modulo 8 (or 128) arithmetic. This number, denoted as the send sequence number, $n(S)$, is contained within the frame header. Information and supervisory frames contain a receive sequence number $n(R)$ to acknowledge the reception of I format frames up to and including the I format frame with send sequence number $n(R) - 1$.[7]

To manage congestion, the maximum number of unacknowledged I format frames for each direction of transmission is restricted to the window size k. The lower window corresponds to the oldest unacknowledged I format frame and the upper window edge is the lower window edge plus k. As acknowledgments are received, the lower window edges are moved and the transmission station can send additional I format frames. The receiving station may acknowledge more than one I format frame.

The transmitting station maintains a copy of each packet it is sending. When the receiving station acknowledges receipt of one or more information frames, the packets associated with those information frames are flushed from the buffer. If the acknowledgment for an I format frame is not received within a specified time, the sending station will query the receiving station with an appropriate supervisory frame. Should the acknowledgment still be missing after n attempts, the transmitter will take alternative recovery actions.

The X.25 packet level procedure (X.25 PLP 1980, 1984, ISO 8208) defines rules to use both logical channels to multiplex user sessions and the link bandwidth more efficiently, in addition to data transfer flow control. The allocation of logical channels can be either static in the case of a permanent virtual circuit (PVC) or dynamic, as in the case in a switched virtual circuit (SVC). X.25 allows up to 4096 logical channels on an individual interface.

Asynchronous and non-X.25 terminals can interface to an X.25 network through a PAD. The PAD performs the necessary X.25 functions on behalf of the asynchronous terminal. CCITT recommendations X.3, X.28, and X.29 define the necessary interface and protocols for the

TABLE 7.2   Standards Describing PADs

X.3 *Packet Assembly / Disassembly Facility (PAD) in a Public Data Network*	X.3 describes the functions of the PAD and defines the operating parameters and the control commands for conversion between serial streams of data with a start-stop terminal and packets with an X.25 terminal.
X.28 *DTE / DCE Interface for a Start-Stop Mode Data Terminal Equipment Accessing the Packet Assembly / Disassembly Facility in a Public Data Network in the Same Country*	X.28 describes the procedures for connection and operation of simple asynchronous DTEs with X.25 packet networks. The PAD performs the function of converting serial streams of information from the asynchronous DTE into X.25 packets for transit in the network.
X.29 *Procedures for Exchange of Control Information and User Data Between a Packet Assembly / Disassembly and a Packet Mode DTE and Another PAD*	X.29 specifies the procedures for an X.25 DTE to control the operation of the PAD during a communication with an asynchronous DTE through the PAD.

support of start-stop terminals (see Table 7.2). X.25 packets that control the PAD are called data-qualified packets and have their Q-bit (bit 8 in the first octet of a packet header) set to 1 by higher-level entities. The Q-bit is always 0 for packets containing user data and all data packets in a complete packet sequence must have the same Q-bit setting. Terminal handling, data forwarding, and other PAD functions associated with asynchronous terminals are controlled by approximately 40 X.29 parameters. Table 7.2 summarizes the PAD standards.

Figure 7.10 depicts an actual network environment. For a public network, for example, one provided by a local Bell Operating Company or US Sprint Data (formerly Telenet), the PAD is located either at the X.25 switch site or at some convenient remote location. For example, a public packet network could consist of 10 switches in the top 10 U.S. cities interconnected with trunks. Then the carrier

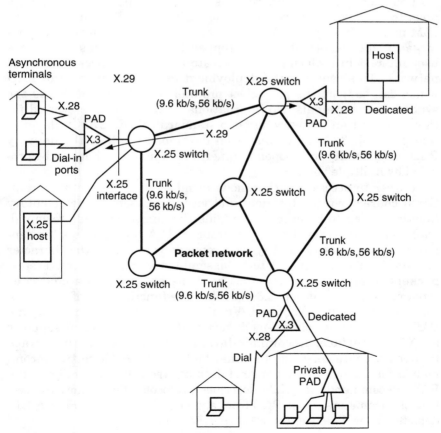

Figure 7.10 PADs in a packet-switched network.

could deploy PADs in the next 25 U.S. cities; these PADs would be connected radially to the closest switch. This topology enables the end users to call a local or regional telephone number and be connected to the packet network without having to incur large connection expenses.

Typically, users can dial into the PAD and obtain the protocol conversion services required to access the packet network and in turn be connected to a host or a database in another part of the country or the world. This works well for users requiring 1 or 2 hours of connection per day. For users needing connection for longer periods of time, a dedicated line from the customer site to the PAD location can be employed (see Fig. 7.10). For companies requiring a large number of connections from a colocated population of users, the PAD can actually be located on the company premises (see also Fig. 7.10). For LAN interconnection applications, the PAD is functionally included in the router.

In business environments there is a large installed base of synchronous terminals, typically for mission-critical applications based on IBM mainframes and SNA. These networks traditionally consist of a large number of private multidropped lines. There has been some user interest in replacing these private lines with a packet-switched network. This requires the deployment of a different type of PAD which can support the poll-select protocol required by SNA and the synchronous terminals. No international standards have evolved for this PAD, but an industry de facto standard called 3270 display system protocol (3270 DSP) has been developed to allow the delivery PAD functionality in an open manner (i.e., a manner that can be provided by multiple vendors).

In order to access a packet-mode terminal or a host from a traditional character-mode terminal and vice versa, some supplementary functions need to be executed outside the character-mode terminal. (Note for comparison that a synchronous SNA terminal is a "screen-mode" device.) One of the key requirements is to enable the character stream from the terminal to be assembled in packets; conversely, packets from the network must be disassembled into a character stream. Recommendation X.3 describes the functionality of the protocol converter that is required. From the network point of view, the PAD is just another packet-mode terminal, which is to say it supports an X.25 interface. Recommendation X.28 defines the interface between a character-mode terminal and the PAD. These two recommendations allow the exchange of data between the terminal and the PAD. Recommendation X.29 describes protocols which allow the setting of parameters in the PAD from a packet-mode device. Figure 7.11 depicts a logical view of the PAD specifications.

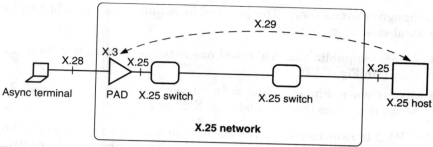

Figure 7.11   Relationship between PAD standards.

The PAD is a protocol converter between the packet-switching network and an asynchronous terminal. The PAD will assemble characters from the character-mode terminal into packet and transmit them to the packet switch. Similarly, it will receive packet from the packet switch and transmit characters to the character-mode terminal. The PAD performs all X.25 protocols functions on behalf of the terminal, such as call establishment, call clearing, and flow control. Recommendation X.3 defines a series of parameters which allow the selection of functions offered by the PAD. Some of the X.3 parameters out of a total of 18 are shown in Table 7.3.[8]

Recommendation X.28 describes the interface between the start-stop terminal and the PAD, and the data exchange and the control

TABLE 7.3   Partial List of X.3 Parameters

Character delete

Escape from data transfer

Echo of characters to terminal (yes, no)

Recognition of data-forwarding signal

Selection of idle timer delay (time-out value which will trigger the sending of a partially full packet)

Ancillary device control (e.g., flow control between PAD and terminal)

Suppression of PAD service signals to terminal (yes, no)

PAD procedure when receiving a break signal from terminal

Discard output

Padding after carriage return

Line folding

Terminal speed (b/s)

Flow control of the PAD by the terminal

Line-feed insertion

Line-feed padding

exchange between them. The physical attachment can be achieved in a number of ways, including:

- Access via public switched telephone network or dedicated lines (as shown in Fig. 7.10)
- Access via public switched data network or leased lines with X-series interfaces (recommendation X.20 or X.20 bis)

The PAD is responsible for the call set-up with the remote entity on behalf of its users. There are two transmission phases: the control phase and the data transfer phase. Control procedures for establishing and terminating the terminal-to-PAD connection are defined in X.28. During the control phase, the PAD does not forward the control data it generates with the local terminal to the packet network. A set of terminal control commands is defined, which must be recognized and processed by the asynchronous terminal. Some of the key commands available to a user are shown in Table 7.4. These commands allow, for example, a virtual call to be established and to be reset or cleared when it is no longer needed. In addition, the X.3 parameters can be examined and changed. Typically, the only command initiated by the start-stop terminal is the command for call setup. Beyond that point, the packet network controls the X.3 parameters via X.29. Whenever a command is sent from the terminal to the PAD, the PAD returns a response in the form of a service signal. An example of a response is "call successfully set up." If the call is not set up or is prematurely cleared, the PAD outputs a "clearing service signal" along with the reason. Some of the X.28 PAD service signals are shown in Table 7.4.

TABLE 7.4   Partial List of Key X.28 Commands and Service Signals

X.28 commands	Call Request
	Call Status Request (STAT)
	Clear Request (CLR)
	Interrupt (INT)
	Interrupt and Discard Data
	Profile Selection (selects sets of PAD parameters)
	Read PAD Parameters (PAR)
	Reset (RESET)
	Set PAD Parameters (SET)
	Set and Read PAD Parameters
X.28 PAD service signals	Call status
	Clear and reason code
	Command acknowledgment
	Error
	Indication of call connected
	Reset and reason code

TABLE 7.5   Partial List of Key X.29 Messages

Set PAD	Sets specified parameters to indicated values
Read PAD	Reads values of specified parameters
Set and Read PAD	Performs the Set PAD and Read PAD function
Parameter indication	List of parameters and values; in response to Read PAD command; from PAD to requester
Invitation to clear	Requests PAD to clear the virtual call
Indication of break	Indication that a break has been received from terminal (this message is to or from the PAD)
Error	Indicates error condition (this message is to or from the PAD)

Once the virtual call is active, there may be a need to exchange control information between the PAD and the remote packet terminal (or PAD). This exchange of control information is described in X.29. The procedures include the reading and setting of PAD parameters by the remote packet device; requesting the PAD to clear the connection; interrupts; and error management. Table 7.5 lists some of the key X.29 messages. The exchange of control information is done through the user data field in the data packets, and not in packet header itself, as one might expect at first. When a remote X.25 device sends a data packet, it must indicate whether the data is to be interpreted by the PAD or sent directly to the terminal. The distinction between real user data and that control information is indicated through the setting of the Q-bit in the packet header, as discussed earlier. Control information is distinguished by having the Q-bit set to 1; user data has the Q-bit set to 0.

Today PADs are basically commodity items offered by a variety of vendors including Codex, Micom, Dynatech Communications, OST, Plantronics Futurecomms, General Datacomm, and Synch Research.[9] Most PADs are available in stand-alone versions; however, the trend is toward integrating "board versions" into other products such as communications controllers, LAN routers, and multiplexers. These board versions are typically software-based and offer the ability to download enhancements as they become available.

PADs typically support between 1 and 2.50 X.25 channels. Over a third of PAD products support between 1 and 16 devices. Another third support up to 32 devices. Many products support more than 64 devices, with some PADs, like Penril Datacomm's VCX-1000, supporting over 250 devices. Cost per device or port ranges between $175 to $250. Most PADs are priced between $1500 and about $6000, with some high-end products in the $50,000 range.

Additional features supported on some PADs could include data encryption, collection of billing statistics, and password protection. There is a continuing strong demand for products that support IBM synchronous protocols. Although the U.S. demand for PADs is relatively flat, the demand for PAD products in Europe should continue to grow in anticipation of the impending deregulation of its telecommunications.

7.3.4 Frame relay service

Frame relay is a recently introduced multiplexed data networking capability supporting connections between user equipment (routers and nodal processors in particular), and between user equipment and carriers' frame relay network equipment (i.e., switches). The traditional WAN approach of connecting a few LANs with routers over dedicated point-to-point lines is no longer adequate in an environment of many remote LANs. Some actual networks can have as many as 1000 routers. LAN managers have sought solutions which reduce the number of dedicated lines in order to keep transmission costs down, while increasing flexibility and making network management easy.

Frame relay service provides interconnection among n sites by requiring only that each site be connected to the "network cloud" via an access line. (Compare this with the $n(n - 1)/2$ end-to-end lines required with dedicated services.) The cloud consists of switching nodes interconnected by trunks used to carry traffic aggregated from many users. See Figs. 7.6 and 7.8(b). In a public frame relay network, the switches and trunks are put in place by a carrier for use by many corporations. In a private frame relay network, the switches and trunks are put in place (typically) by the corporate communications department of the company in question.[6] Each approach has advantages and disadvantages within the framework of a corporate network. In the sequel the term *frame relay* refers generically to either the service or to the supporting technology, depending on the context; usually *frame relay service* refers to a public carrier service, while *frame relay technology* implies platforms for private network solutions. Over a dozen U.S. carriers now offer the service, including the Bell Operating Companies, and the number is likely to grow.[5,6,10–12]

The frame relay protocol supports data transmission over a connection-oriented path; it enables the transmission of variable length data units, typically up to 4096 octets (in some cases as high as 8192 octets) over an assigned virtual connection (see Fig. 7.12). As is the case in X.25, frame relay standards specify the user interface to a device or network supporting the service. This interface is called frame relay interface (FRI). An FRI supports access speeds of 56 kb/s, $n \times 64$ kb/s and 1.544 Mb/s (2.048 Mb/s in Europe). Some vendors are attempting to extend the speed to 45 Mb/s. The service can be

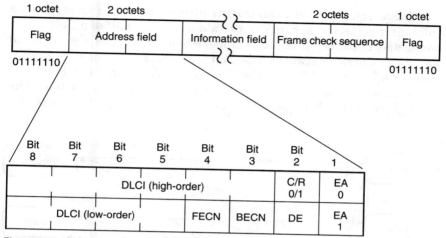

Figure 7.12 A frame relay frame. C/R: Bit intended to support a command/response indication. The use of this field is application-specific. EA: Address field extension bit. DE: Discard eligibility indicator. BECN: Backward explicit congestion notification. FECN: Forward explicit congestion notification. DLCI: Data link connection identifier.

deployed (1) in a point-to-point link fashion between two routers, (2) using customer-owned frame relay nodal processors (basically, frame relay switches, but, typically, using cell relay trunking), and (3) utilizing a carrier-provided service.

In order to save money, frame relay networks rely on the statistical variation of the traffic input from the pool of users to provide nominal combined input exceeding the actual throughput of the trunks in the backbone network. This is the reason why frame relay networks work best for bursty traffic applications. Because of the statistical nature of the network, carriers typically tariff a "committed" information rate (CIR), which, supposedly, the network can carry from a specified source or sources. In actuality, the network may become congested and data submitted to the network, at the CIR or even below, *may or may not be delivered.*

In a frame relay environment the user needs a device (such as a router) to run the frame relay protocol, rather than the X.25 protocol. Some networks provide conversion for nonframe relay devices. The network forwards frames rather than X.25 packets. The transit delay through the network is about an order of magnitude lower because the frame relay protocol is simpler. Frame relay provides error detection but not correction (which was provided by X.25); the correction must be supported by the user equipment. Frame relay as currently implemented supports a "connection-oriented" permanent virtual circuit service (as does X.25); in the future it may also support a switched virtual circuit service (as does X.25).[5,6,10–12]

A frame relay network can cost as little as one-third as much as a fully interconnected network of private lines (one-half of a fully interconnected network, when also considering the cost of the required switches). Carrier networks based on frame relay provide communications at up to 1.544 Mb/s (in the United States), shared bandwidth on demand, and multiple user sessions over a single access line. The throughput is much higher than that available for packet switching

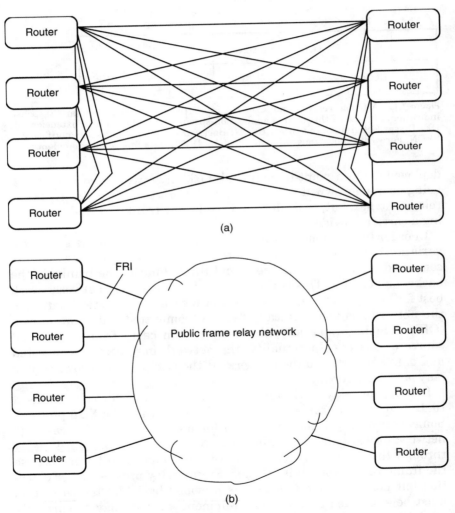

Figure 7.13 Clear advantage of public frame relay compared to use of point-to-point lines. (*a*) Without public frame relay service: 28 point-to-point links. (*b*) With public frame relay service: 8 links.

(refer to Fig. 7.2) and is adequate for today's LAN applications (discussed in Chap. 2; data-intensive applications are better off with SMDS). A carrier can multiplex the traffic of one user with that of other users and can, therefore, pass back to the users the economic advantages of bandwidth sharing. Without carriers or private switches, dedicated T1 links between two sites to be interconnected are needed, regardless of the protocol used over the link. Figure 7.13 depicts this scenario graphically.

Another way to benefit from frame relay is to use it in conjunction with a frame relay nodal processor (also known as "fast packet switch") (see Fig. 7.14). Using nodal processors can, in many instances, be cost-effective, since the user can obtain from the backbone bandwidth-on-demand rather than on a preallocated (and inefficient) basis. The "saved" bandwidth is then available to other users of the same backbone, in theory minimizing the amount of new raw bandwidth the corporation needs to acquire from a carrier in the form of additional T1 or fractional T1 links. These devices typically implement the frame relay protocol on the user's side and cell relay in the trunk side. Cell relay principles facilitate dynamic bandwidth allocation. In the private network application, the user leases from a carrier private lines between the remote devices and the nodal processors, and between the nodal processors. The user employs frame relay to statistically multiplex traffic in a standardized way, in order to achieve better utilization of the (now common) transmission resources. The nodal processors must be housed in selected user locations. About half a dozen vendors now manufacture nodal processors.

Two frame relay services are possible: a switched implementation in conjunction with ISDN, using the recent CCITT Q.933 protocol for call setup, and a PVC implementation. The PVC does not require call setup and call termination, but it is obviously not as efficient in resource utilization as SVC. All current public network services are PVC-based; user-owned nodal processors also only support a PVC implementation. SVC frame relay service should become available in 1994. The initial commercial impetus in the United States has been toward the deployment of frame relay service within the context of a private corporate network. More recently, a number of carriers have started to offer the service as a public offering.

Like X.25, frame relay specifies the interface between customer equipment and the network (i.e., the UNI), whether the network is public or private. Development of standards for frame relay started in 1986; work accelerated in 1989, after the publication of the original CCITT frame relay standards. Frame relay standards were published in final form in 1991. Three key ANSI standards are T1.606-1990, T1.617-1991, and T1.618-1991. T1.606-1990 specifies a framework for

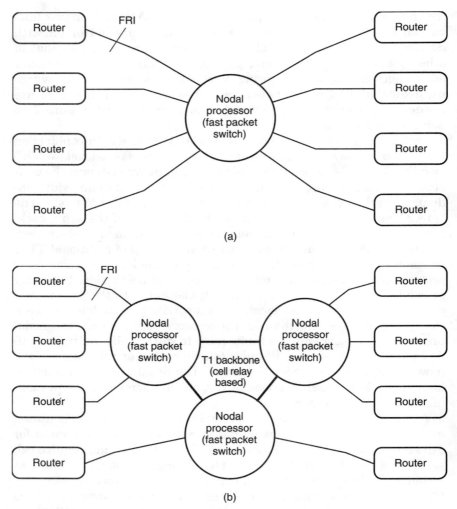

Figure 7.14 Private frame relay configurations. (*a*) Private frame relay node consisting of one nodal processor. (*b*) Private frame relay node consisting of multiple nodal processors, interconnected by cell relay trunks.

frame relay service in terms of user-network interface requirements and internetworking requirements. T1.617-1991, Annex D, specifies critical network management functions, particularly useful in the public frame relay service context. The protocol needed to support frame relay is defined in T1.618-1991 (LAPF Core); the protocol operates at the lowest sublayer of the data link layer and is based on the core subset of T1.602 (LAPD). The frame relay data transfer protocol defined in T1.618/LAPF Core is intended to support multiple simultaneous end-user protocols within a single physical channel. This protocol provides transparent transfer of user data and does not restrict the contents, format, or coding of the information or interpret the structure.[5,6,10–12]

Frame relay allows users of multiple routers supporting traditional LANs to connect them in an effective manner. Three basic ways of using frame relay to connect LAN are as follows:

1. Deploy a private frame relay network using frame relay nodal processors. Instead of physical point-to-point links, this approach only requires connecting the routers to the nodal processors with a single physical link. Connection between various routers is accomplished with PVCs, which are set up by the LAN administrator.

2. Use a PVC-based carrier-provided frame relay network. Instead of many physical point-to-point links, this approach only requires connecting the routers to the carrier's switch with a single physical link. Connection between various routers is accomplished with PVCs which are established at service subscription with the carrier.

3. Use a SVC-based carrier-provided frame relay network. Connection between various routers is accomplished as needed by establishing a real-time SVC which is in existence only for the duration of the session. (This is a solution of the future.)

The financial advantage of a frame relay network becomes more substantial when the number of routers is high (half a dozen to a dozen, or more) and when the distances between routers is considerable (hundreds or thousands of miles). If the routers are all located within a small geographic area, like a city, a county, or a LATA, the economic advantage of eliminating lines is less conspicuous. Table 7.6 summarizes some of the benefits of frame relay.

TABLE 7.6 Some Benefits of Frame Relay

Private	Port and link sharing
	Bandwidth-on-demand
	High throughput and low delay
	Ease of network expansion
	Ease of transition from existing network
	Cohesiveness and symbiosis with LANs
	Simplified network administration
	Standards-based
	Economic advantages
Public	Major reduction in transmission costs
	Low start-up cost
	Ability to support a variety of user equipment
	Ability to transmit instantaneous bursts exceeding the throughput class

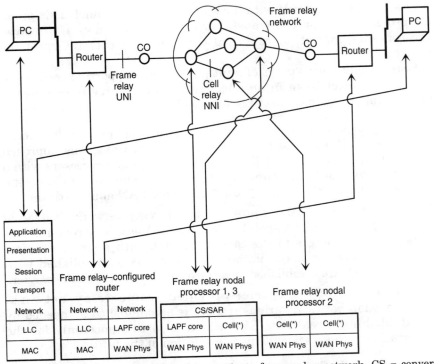

Figure 7.15 Protocol stacks required in a private frame relay network. CS = convergence sublayer; SAR = segmentation and reassembly; * = ATM or vendor-proprietary.

Figure 7.15 depicts a typical frame relay network protocol architecture. These stacks must be implemented in the user equipment and in the nodal processors. In the example, the LANs access the frame relay node via routers configured to terminate the frame relay interface. Three network nodes have been provisioned to logically interconnect the end-user equipment via permanent virtual circuits. Node 1 and node 3 terminate the end-user equipment directly over link with a frame relay interface. On the trunk side, they support cell relay functions. These are "invisible" to the user but accomplish the dynamic bandwidth sharing function.

7.3.5 Switched multimegabit data service

SMDS is similar in some respects to frame relay service in that it provides a "cloud" of switches and trunks, eliminating the need for dedicated point-to-point links (see Fig. 7.8). However, there are some substantial differences as follows.[12,13]

1. SMDS operates at 1.544 Mb/s (1.17 Mb/s actual) and 44.736 Mb/s (34 Mb/s actual); hence it is better suited for the evolving LAN

applications such as those listed in Table 2.7.* It is a connection-less service, while frame relay is connection-oriented.

2. The user needs a device (such as a router) to run the SMDS inter-face protocol (SIP).

3. The transit delay is smaller than that experienced in a frame relay network (guaranteed to 20 ms 95 percent of the time with T3 access), and is tightly controlled by the network, as are other grade-of-service parameters (lost cells, misrouted cells, etc.). This is far from being the case in a frame relay network.

4. SMDS is a robust carrier-grade service. Sophisticated manage-ment and operations features are built into SMDS to facilitate its administration and monitor service quality. Security is also guar-anteed.

5. Traffic can be input at the T1 or T3 rate (as well as other rates in between). SMDS may support synchronous optical network (SONET) rates in the future.

6. SMDS is a "connectionless" service; that is, no call setup or tear-down is needed, making it faster.

SMDS should be widely available from the Bell Operating Companies in 1993.

7.3.6   Cell relay service

As shown in Fig. 7.2, cell relay is an important ultra-high-speed ser-vice now beginning to appear. The reader is referred to Chap. 9.

References

1. D. Minoli, "Internetworking LANs: Repeaters, Bridges, Routers and Gateways," *Network Computing,* October 1990, pp. 96 ff.
2. D. Minoli, "Connecting LANs to WANs, Low-Speed, Non-Switched Solutions," *Network Computing,* November 1990, pp. 86 ff.
3. D. Minoli, "Interconnecting LANS over a Wide Area: High-Speed, Nonswitched Solutions," *Network Computing,* December 1990, pp. 82 ff.
4. D. Minoli, "Interconnecting LANS over a Wide Area: Switched Solutions," *Network Computing,* February 1991, pp. 81 ff.
5. A. Tumolillo, *Frame Relay vs. SMDS vs. T1: The Best Technology/Service Fit for Networked Applications,* Probe Research, Cedar Knolls, N.J., 1992.
6. *The Frame Relay Alternative: A Network Manager's Guide to Understanding, Evaluating, and Implementing a Private Frame Relay Network,* Document 55-300-915, AT&T Network Systems, 1992.
7. D. Minoli, *Telecommunication Technology Handbook,* Artech House, Norwood, Mass., 1991.

---

*There is also work under way for access at $n \times 64$ kb/s.[14]

8. A. Meijer and P. Peeters, *Computer Network Architectures,* Computer Science Press, Rockville, Md., 1982.
9. Data Networking, Tab 3406 Packet-Switching Equipment, DATAPRO, Delran, N.J., 1991.
10. D. Minoli, "Technology Overview: Frame Relay," Datapro Report CA09-020-501, January 1991.
11. D. Minoli, "The New Wide Area Networking Technologies: Frame Relay," *Network Computing,* May 1991, pp. 102 ff.
12. D. Minoli, *Enterprise Networking, Fractional T1 to SONET, Frame Relay to B-ISDN,* Artech House, Norwood, Mass., 1993.
13. D. Minoli, "The New Wide Area Technologies: SMDS and B-ISDN," *Network Computing,* August 1991, pp. 88 ff.
14. M. Strizich, "Low-Speed SMDS in Focus," *Communications Week,* November 30, 1992, pp. 21 ff.

8

Second-Generation LANs: Fiber-distributed Data Interface

8.1 Introduction

The fiber-distributed data interface (FDDI) is a set of standards that defines a 100 Mb/s shared-medium LAN utilizing fiber (single-mode and multimode) and twisted-pair cabling. Standardization work started in 1982 and reached stability in the late 1980s, with specs related to management slower in attaining agreement and maturity. FDDI uses a token-based discipline and was developed to support backbone interconnection of LANs within a building or campus. Logically, it consists of a dual ring (although it may be implemented as a physical star).[1,2] As shown in Fig. 3.2, the FDDI encompasses both the MAC and physical layers and interfaces with the IEEE 802.3 LLC.

Why do we need a second-generation LAN? Network planners and analysts see the potential for severe network bottlenecks occurring now and in the near future; some of the reasons for the network congestion can be attributed to factors listed in Table 8.1.[3]

Proponents see FDDI as a "multivendor local area network standard offering an industry-standard solution for organizations that need flexible, robust, high-performance networks."[3] FDDI, however, has seen slow penetration because the cost of the device attachment started out high and went down only slowly (starting at $10,000 per device, down to $5000 in the late 1980s, $2000 in the early 1990s, and as little as $975 at time of publication).

FDDI supports data transmission. A second set of standards, known as FDDI-II is aimed at isochronous traffic, such as voice and compressed digital video at 1.2 to 4 Mb/s. Work on FDDI-II is still

TABLE 8.1   Some Recognized Reasons for the Need for Second-Generation LANs

Increasing number of users being added to the network

Increasing computing power of smaller desktop systems

Increasing business needs for more complex networks spanning longer distances

Increasing traffic loads on existing backbone networks

Increasing number of client-server facilities

Increasing deployment of a new, high-speed applications

Increasing number of window terminals

Increasing requirement for networks to support graphics-intensive applications

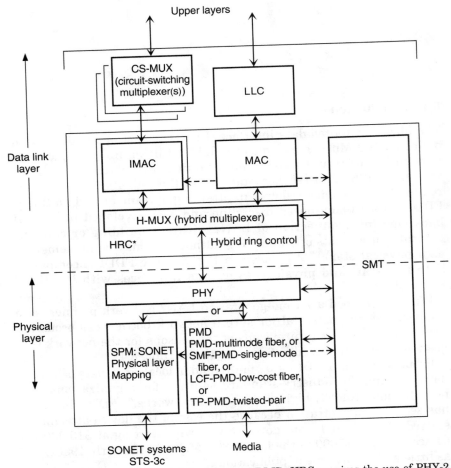

Figure 8.1   FDDI reference model (including FDDI-II). HRC requires the use of PHY-2 and MAC-2; otherwise, any combination of MAC or MAC-2 and PHY or PHY-2 is allowed. * = optional.

TABLE 8.2   FDDI Highlights

Shared medium using a token-passing MAC scheme similar (but not identical) to the IEEE 802.5 token ring standard.

Compatibility with IEEE 802 LANs by use of the 802.2 LLC.

Ability to use a variety of physical media, including multimode fiber, single-mode fiber, and twisted pairs (for both UTP and STP under development).

Distributed clocking to support a large number of devices on the ring.

Sophisticated encoding techniques to ensure data integrity.

A dual-ring topology for fault tolerance.

Operation at a data rate of 100 Mb/s (signaling at 125 Mbaud) and the ability to sustain an effective data transfer rate of 80 Mb/s. The averaged per capita bandwidth of a fully configured network (500 FDDI devices) is 200 kb/s (100,000,000/500).

Support for 1000 physical attachments (500 devices).

FDDI backbone configuration supports wiring topologies defined EIA/TIA568 standard for commercial building wiring.

A maximum fiber length of 200 km (124 mi) with a maximum of 2 km (1.2 mi) between adjacent devices.*

Ability to dynamically allocate bandwidth so that both "synchronous" and "asynchronous" services can be provided simultaneously.†

Concentrators are key building blocks of FDDI network; attach directly to the dual FDDI ring; connect single-attachment stations and dual-attachment stations to the primary ring.

Dual-attachment stations connect to both rings or to the primary ring through a concentrator; single-attachment stations connect to the primary ring through a concentrator.

---

*FDDI limits total fiber length to 200 km (124 mi); since the dual-ring topology effectively doubles medium length in the event of a ring wrap, the actual length of each ring is limited to 100 km (62 mi).

†The terms *synchronous* and *asynchronous* are used in FDDI with special meanings that relate to ring transmission. Asynchronous transmission is a method of communication in which information is sent when the token holding rules allow transmission, on a statistical basis. In synchronous transmission information is guaranteed a percentage of the ring's bandwidth. There is no relationship between this use of "asynchronous" and B-ISDN's asynchronous transfer mode.

opment of FDDI-II standards. The primary standard is the hybrid ring control (HRC). HRC defines additional MAC procedures above and beyond FDDI (now also retroactively called FDDI-I) to support voice, real-time data, and other isochronous services requiring circuit-switched services over an FDDI network. MAC-2 addresses data link issues when the MAC is used with the HRC. As seen in Fig. 8.1, the isochronous MAC (IMAC) provides the interface between the HRC and the circuit-switched services, while the hybrid multiplexer provides the mechanism so that both circuit-switched and packet-switched data can be carried simultaneously. PHY-2 addresses additional physical layer issues for FDDI-II. Actual FDDI-II products of any consequence are yet to emerge, and the commercial outlook of this technology is uncertain (particularly with the appearance of more powerful ATM-based LANs discussed in Chap. 9).

8.2   Physical Aspects

A directly attached FDDI *node* is an active element on the network that is capable of repeating incoming transmissions but is not necessarily capable of performing data link layer error recovery functions. Thus, it contains at least one PHY and PMD entity and zero or more MAC entities.[6] A directly attached FDDI *device* is an addressable node on the network. It must contain at least one PHY, PMD, and MAC entity. All FDDI devices are nodes, but not all nodes are devices. In the sequel the more general term *device* is employed.

8.2.1   FDDI devices

FDDI allows interconnection of up to 500 devices at a shared-speed of 100 Mb/s over distances reaching 100 km. Devices connect to the FDDI network either directly or through a concentrator, so named because it concentrates, or gathers together, several lines in one central location. The use of concentrators maintains ring integrity even when one or all devices attached to it are off-line (i.e., powered down). This means that users can power down the attached devices without disrupting the FDDI network.[3] Also they are used to reduce the per terminal connection cost.

FDDI defines four types of attachment devices (stations) in order to allow for the deployment of complex topologies and for designs with adequate reliability. The devices types are:

1. Dual-attachment stations (DASs)

2. Single-attachment stations (SASs)

3. Dual-attachment concentrators (DACs)

4. Single-attachment concentrators (SACs)

To enhance the reliability of an FDDI network, the standard provides for a dual-ring implementation. SASs attach only to the primary ring; in case of ring failure, a SAS may be disconnected from the network. DASs are connected to their neighbors by two links that transmit in opposite directions. (One ring is called the primary ring and the other the secondary ring.) When a link failure occurs, the devices on either side of the link reconfigure. This isolates the fault and restores a continuous ring. A dual-attachment concentrator is connected to both the primary and secondary rings and is fault-tolerant to ring failures. A single-attachment concentrator is used to connect SAS within a logical tree. Table 8.3 shows the nomenclature

TABLE 8.3 Port Types Used in FDDI

Port A	Connects to the incoming primary ring and the outgoing secondary ring of the FDDI dual ring. This port is part of a DAS or a DAC.
Port B	Connects to the outgoing primary ring and the incoming secondary ring of the FDDI dual ring. This port is part of a DAS or a DAC and is also used to connect a DAS to a DAC.
Port M	Connects a concentrator to an SAS, DAS, or another concentrator (DAC or SAC); only implemented in a concentrator.
Port S	Connects an SAS or an SAC to a concentrator (DAC, SAC).

used to describe device ports. Figure 8.3 depicts the functional internals of a concentrator.

The most general FDDI topology is a dual *ring of trees*. Here a dual ring can serve as the backbone ring for a group of other FDDI networks. This is also known as a *trunk ring*. Figure 8.4 illustrates an example that shows the use of all four device types. The backbone is a dual ring consisting only of DASs and DACs, i.e., devices that are capable of supporting attachment to two rings. DACs support a number of end-user devices, SACs, or both. (Each DAC becomes the root of a tree.) Note that the main functions of the secondary path are to aid in ring initialization and reconfiguration and to provide backup to the primary ring. Under normal conditions, the backup ring is idle; that is, no user information is transmitted over it.

SASs attach to the DAC by means of a single ring. In the event of a failure of the SAS or its connection to the concentrator, the concentrator will isolate the SAS. Therefore, the continuity of the backbone ring is maintained. To achieve a tree structure of depth greater than two, SACs may be cascaded. A SAC may attach to a DAC or another SAC, and may support one or more SASs.

It should be noted that even with a complicated tree structure, an

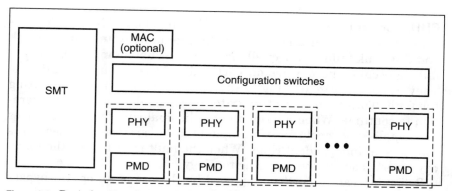

Figure 8.3 Basic functional view of a concentrator.

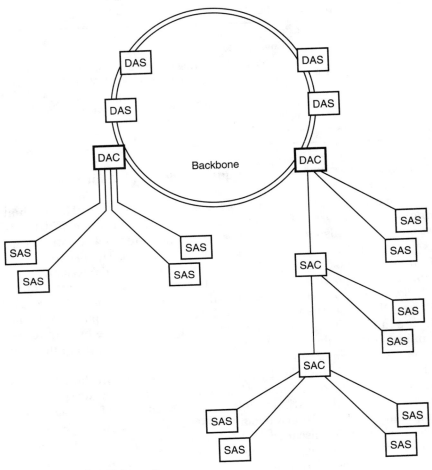

Figure 8.4   FDDI dual ring of trees.

FDDI configuration still remains a ring topology. Figure 8.5 illustrates this fact, even under various link failures. Note that SAS-link and SAC-link failures affect all leaves of the subtending tree. When a failure occurs, the devices on each side of the failure reconfigure. They wrap the primary ring to the secondary ring, thereby isolating the fault. This restores continuity to the ring, allowing normal operation to continue. When a wrap occurs, the dual-ring topology changes to a single-ring topology. If multiple faults occur, the ring segments into several independent rings. When the fault is corrected, the topology reverts back to the dual ring topology.[3]

Reliability can be further improved by the use of an optical bypass

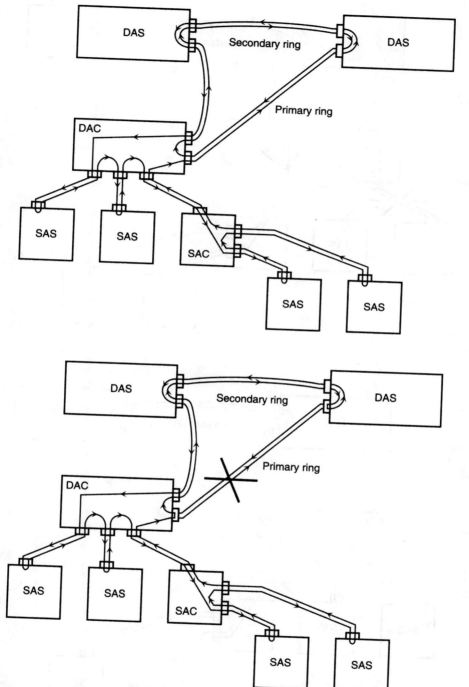

Figure 8.5a FDDI reconfiguration: (*top*) normal operation; (*bottom*) backbone ring failure.

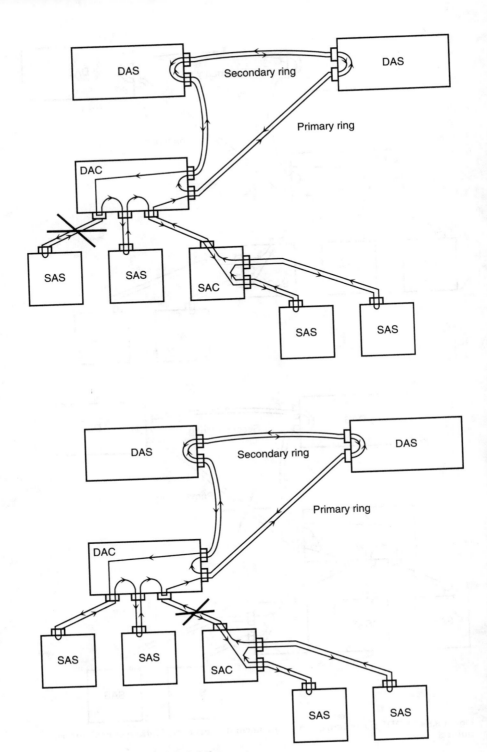

Figure 8.5b SAS and SAC link failures.

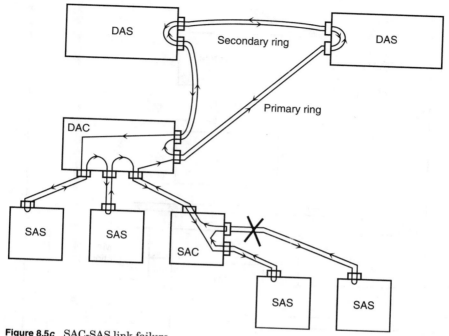

Figure 8.5c SAC-SAS link failure.

switch, which can be optionally installed in DACs or DASs (see Fig. 8.6). The optical switch bypasses the device's active components supporting the receiver and transmitter functions so that the optical signal from the device in question is passed directly to the next device. Bypassing can be activated automatically by the device itself, a neighbor device, or an administrator. However, optical bypass switches introduce attenuation, thereby impacting the overall power budget; hence, only a limited number of devices in the dual ring can be equipped with such a capability. Figure 8.7 depicts the protocol architecture in a number of FDDI devices with and without an optical switch.

8.2.2 FDDI environments and network topologies

Four application environments have been described in the standards for FDDI networks. These user environments differ primarily in the number of devices attached to the network and the geographic size of the network.[6]

1. *Data center environment.* This back-end application is characterized by a relatively small number of devices (a few dozen), such as mainframes, high-speed peripherals, and servers. High through-

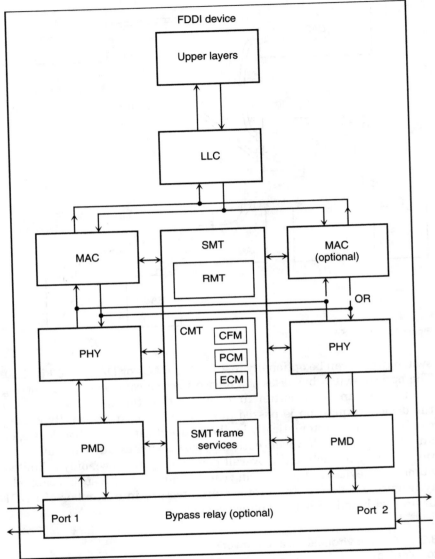

Figure 8.6   Optional bypass relay in a dual-attachment device.

put and reliability-fault tolerance are essential (most devices are dual-attached devices). This environment assumes a fiber length of no more than about 400 m between adjacent devices and a total ring size of no more than about 20 km.

2. *Office or building environment.*   This front-end application is characterized by a relatively large number of single-attachment

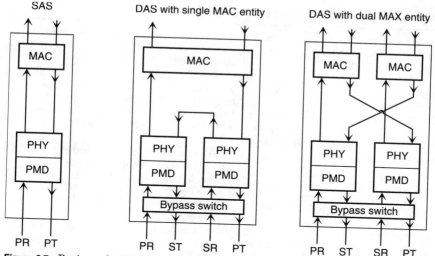

Figure 8.7 Basic mode FDDI protocol architecture for SAS and DASs. PR = primary ring receive; PT = primary ring transmit; SR = secondary ring receive; ST = secondary ring transmit.

devices connected with a star wiring scheme. The devices in this application typically consist of terminal servers and concentrators, workstations, PCs, and MPCs. Star wiring is supported by the use of concentrators. Since wiring concentrators are never powered down; they maintain the integrity of the network when individual user devices are turned off. As it will be seen later, each device directly connected to an FDDI network needs to actively retransmit every bit it receives.

3. *Campus backbone network environment.* This is characterized by devices located in multiple buildings, with links up to 2 km. It is assumed that the user has right of way.

4. *Multicampus environment.* This is characterized by clusters of devices located at different sites, possibly separated by distances of 60 km or more. A multicampus environment may typically have the requirement to cross rights of way owned by a variety of agencies.

To date the office or building environment and the campus backbone network environment are the more common candidates for FDDI. From an implementation perspective, some topologies are more common than others. Table 8.4 lists some of these. (See Fig. 8.8.)

8.3 FDDI Protocols

Table 8.5 provides a summary of the key FDDI standards. Table 8.6 lists other (newer) standards or draft standards (as of August 1992). There

TABLE 8.4   Typical FDDI Topologies

Stand-alone concentrator with attached devices	Single concentrator and attached devices. Typical for connection of multiple high-end workstations in a workgroup. Also typical for interconnection of multiple LANs. (Here each FDDI device is a bridge.)
Dual ring	Set of DASs connected to form a single dual ring. Typical for situations with a limited number of users. Also typical for interconnection of interdepartmental LANs. (Here each FDDI device is a bridge.)
Tree of concentrators	Concentrators are wired in a hierarchical star arrangement with one concentrator serving as the root of the tree. Typical for interconnecting large groups of users devices. This topology facilitates adding and removing concentrators and devices without disrupting users.
Dual ring of trees	The most sophisticated topology. Critical devices can be incorporated into the dual ring for maximum availability while less critical devices are connected as leaves.

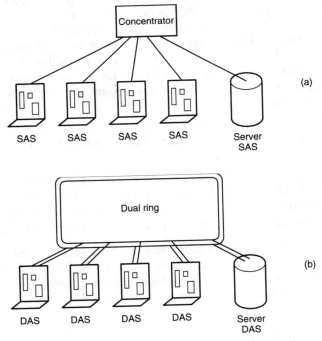

Figure 8.8 Some examples of actual FDDI networks in the workplace. (*a*) Small workgroup standalone concentrator network. (*b*) Small workgroup dual ring design (high availability needed).

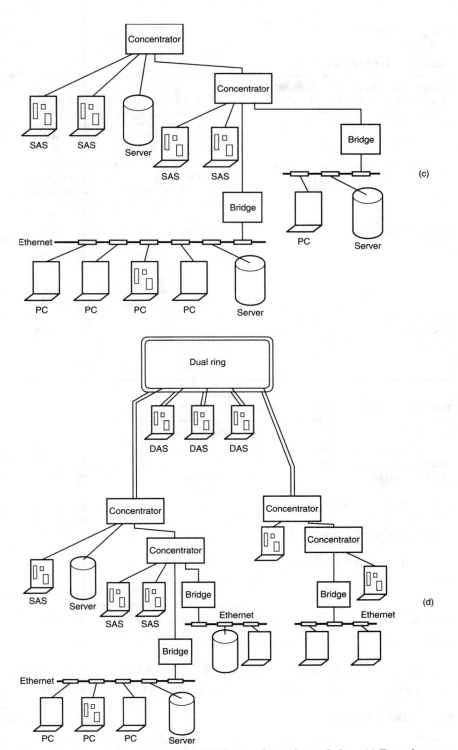

Figure 8.8 Some examples of actual FDDI networks in the workplace. (c) Tree of concentrators for large workgroups. (d) Dual ring of trees, typical of campus applications.

TABLE 8.5   Key FDDI Standards

ANSI X3.166-1990/ISO 9314-3:1990 *Physical Layer Medium-Dependent* (*PMD*)	This standard corresponds to the lower portion of the OSI physical layer. PMD defines the transmit-receive power levels, optical transmitter and receiver interface requirements, and cable and connector specifications.
ANSI X3.148-1988/ISO 9314-1:1989 *Physical Layer Protocol* (*PHY*)	This medium-independent standard corresponds to the upper portion of the physical layer. PHY defines line states, symbols, encoding-decoding techniques, clocking requirements, and framing requirements.
ANSI X3.139-1987/ISO 9314-2:1989 *MAC*	This standard corresponds to the lower portion of the OSI data link layer. MAC defines data link addressing, frame formatting and checking, medium access, error detecting, and token handling. Its major function is to deliver formatted data to nodes attached to the FDDI LAN
SMT-LBC-177 (letter ballot comments)/ ISO WD (working document) 9414-6 *Station Management* (*SMT*)	SMT defines the system management services for the FDDI protocols. SMT includes facilities for connection management, node configuration, recovery from error conditions, and the encoding of SMT frames.

TABLE 8.6   Additional FDDI Standards

MAC-2, Revision 4 (10/29/90)	Enhanced Medium Access Control.
PHY-2, Revision 4.1 (3/5/91)	Enhanced Physical Layer Protocol.
HRC, Revision 6.2 (5/14/91) ISO DIS 9314-5	Hybrid Ring Control. X3.186-199x: X3 Letter Ballot. Previous corrections made; in public review at press time.
SMF-PMD, Revision 4.2 (5/18/90) ISO CD 9314-4	Single-mode Fiber PMD. X3.184-199x approved; editing and publication under way at press time.
CTPICS, Revision 2.2 (2/18/92)	Conformance Test PICS Proforma for FDDI. Under development.
LCF-PMD, Revision 0.2 (2/18/92)	Low cost fiber PMD. Under development.
SPM	SONET Physical Layer Mapping. No document available. Mapping has been established. Other aspects under development.
TP-PMD	Twisted-pair (copper) PMD. Under development. MLT-3 being finalized.

are over a dozen standards supporting FDDI and FDDI-II. Figure 8.9 shows a summary of the basic primitives required for FDDI operation.

8.3.1   FDDI MAC protocol

The MAC protocol corresponds to the lower sublayer of the OSI data link layer. The MAC standard defines rules for medium access,

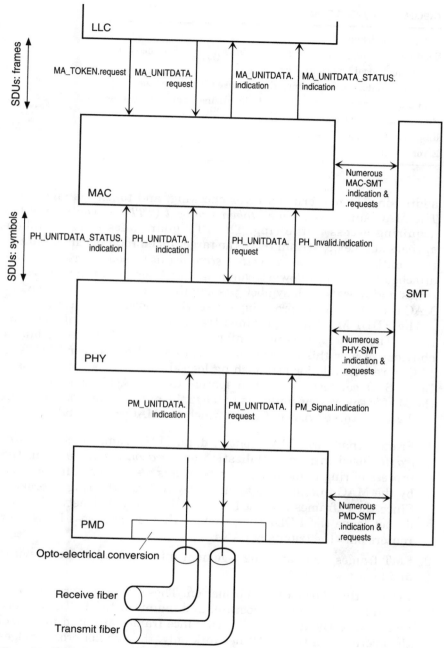

Figure 8.9 Basic protocol primitives.

TABLE 8.7 Key MAC Functions

Constructing frames and tokens

Sending, receiving, repeating, and removing frames from the ring

Delivering logical link control frames

Fair and equal access to the ring through use of the timed token

Communications between attached devices using frames and tokens

Ring initialization ("claiming")

Ring fault isolation ("beaconing")

Error detection

---

addressing, frame formats, error checking, and token management. The MAC supports a *timed-token protocol* (*TTP*) as the means for acquiring access to the ring. The TTP guarantees that the token appears at a device within twice the target token rotation time. Data is transmitted serially as a *symbol stream* (discussed later) from one attached device to its downstream neighbor. Each device in turn *regenerates and repeats* each symbol, passing the symbol to the next device. MAC is responsible for providing the services shown in Table 8.7.

The FDDI MAC also provides transparent services to the LLC layer. The LLC passes user information to the MAC layer, but is oblivious of how this information is transferred to its peer LLC. The LLC is responsible for establishing logical connections between peer LLCs. Such connections are transparent to the MAC in the sense that the MAC's sole responsibility is to move the bits, not interpret them.

There are three types of frames which the MAC must handle:

1. Frames that carry MAC control data. MAC frames include *claim frames* used in ring initialization and *beacon frames* used in the process of ring fault isolation; a beacon is a specialized frame used by the MAC to announce to other devices that the ring is broken. Since these frames are used to initialize the FDDI network, they do not leave the FDDI ring; that is, they do not cross bridges or routers onto the tributary LANs.

2. SMT frames that carry management information between devices and layers.

3. Frames that carry LLC information. These frames carry information that helps control, operate, and maintain the FDDI network and its directly attached devices. Since frames are used to control the operation of the FDDI network only, they do not cross bridges or routers onto the tributary LANs.

A token is a unique symbol sequence that circulates around the ring. A device (i.e., front-end workstation, concentrator, or bridging device) wishing to transmit must wait until it detects a token passing by. The device then captures the token by interrupting the token transmission (not repeating the token onto the network). After the "captured" token is completely received, the device begins transmitting its own LLC frames, that is, PDUs. The LLC frames make a round trip past all the other devices on the ring and are finally purged by the transmitting device.

When the token is captured by a device, other devices wishing to transmit must wait. The device which has captured the token transmits one or more frames for as long as the token holding rules allow and then reissues the token onto the ring. That is, it reissues the token when it has completed transmission or exhausted its available transmission time. When the device that captured the token inserts a new token on the ring, it enables other users to obtain bandwidth. The MAC supports both synchronous and asynchronous communication (see Table 8.1). In asynchronous data transmission, all directly attached devices are allocated a transmission time based on the target token rotation time, the time it should take for a token to rotate around the ring.

Figure 8.10 provides an example of the FDDI token operation. In this figure, device A seizes the token; at this point it transmits a frame [Frm A → C]. Next it immediately releases (transmits) the token. Frame [Frm A → C] is addressed to device C, which copies it as it comes by over the ring. The frame eventually returns to A, which absorbs it by taking it off the ring. Meanwhile, device B could have seized the token issued by A and transmit a frame [Frm B → D] for device D, followed by a token. This action could be repeated a number of times, so that at any one time, there may be multiple frames occupying the ring, but in a noninterfering manner. Each device is responsible for removing its own frames based on the source address field.

Figure 8.11 depicts the FDDI MAC frame format. FDDI defines the contents of this format in terms of symbols. Each symbol can be viewed as a nibble (4 bits, also known as a data quartet) or as a 5-bit code. Data is encoded 4 bits at a time and sent 5 bits at a time; this is called 4B/5B encoding. The MAC layer receives standard user octets (plus PCI) from the LLC layer and transmits user octets (plus PCI) to the LLC layer. Logical symbols are used in communicating with the PHY layer. The PHY then translates the logical 5-bit nonreturn-to-zero code into a transmission-ready 5-bit nonreturn-to-zero inverted signal. This topic is further discussed below. The point to keep in mind at this juncture is that 5-bit symbols are given by the MAC to PHY and 5-bit symbols are received by the MAC from PHY.

In the basic mode (FDDI-I), a frame is limited to 9000 symbols, that

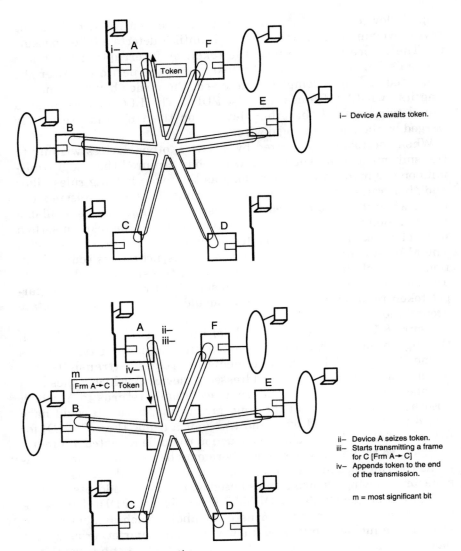

Figure 8.10*a*   FDDI token operation.

is, 4500 octets; in the hybrid mode (FDDI-II), frames are limited to 17,200 symbols, that is, 8600 octets.

The MAC frame fields are as follows.

Preamble.   This field synchronizes the frame with each device's clock. The originator of the frame uses a field of 16 idle (I) symbols (64 bits); subsequent repeating devices may change the length of the field con-

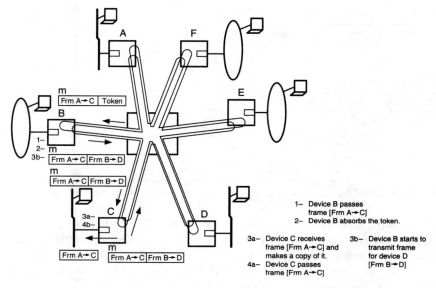

1– Device B passes
frame [Frm A→C]
2– Device B absorbs the token.

3a– Device C receives
frame [Frm A→C] and
makes a copy of it.
4a– Device C passes
frame [Frm A→C]

3b– Device B starts to
transmit frame
for device D
[Frm B→D]

4b– Device C passes [Frm B→D]

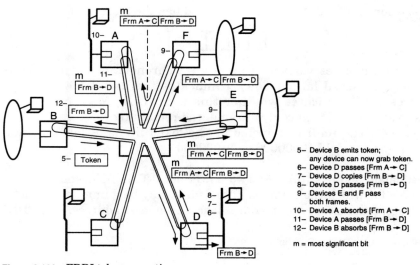

5– Device B emits token;
any device can now grab token.
6– Device D passes [Frm A→C]
7– Device D copies [Frm B→D]
8– Device D passes [Frm B→D]
9– Devices E and F pass
both frames.
10– Device A absorbs [Frm A→C]
11– Device A passes [Frm B→D]
12– Device B absorbs [Frm B→D]

m = most significant bit

Figure 8.10*b* FDDI token operation.

FDDI token frame

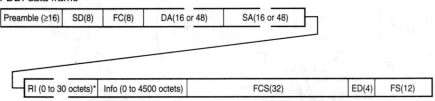

Figure 8.11   FDDI frames (fields shown in bits unless noted otherwise). * = MAC-2.

sistent with local clocking requirements. The idle symbol I is a nondata fill pattern. The actual form of a nondata symbol depends on the signal encoding on the medium. See Table 8.8 for typical line status symbols (although the actual 4B/5B signal encoding is done at the PHY layer, the symbols are generated at the MAC layer).

Starting delimiter (SD; 8 MAC bits).   This field indicates the start of the frame. It is coded as JK (see Table 8.9; J and K are nondata symbols).

Frame control (FC; 8 MAC bits).   This field has the bit format

where c indicates whether this is a synchronous or asynchronous frame (discussed later below); l indicates the use of 16- or 48-bit addresses; ff indicates whether this is an LLC, MAC control, or reserved frame. For a control frame, the remaining zzzz bits indicate the *type* of control frame, including the token frame. A token is indicated by a FC field 10000000 or 11000000.

Destination address (DA; 16 or 48 MAC bits).   This field specifies the devices for which the frame is intended. It may be a unique (physical) address, a multicast group address, or a broadcast address. The FDDI

TABLE 8.8   Line Status Symbols

Function or 4-bit group (4B)	Group code (5B)	Symbol
Quiet	00000	Q
Idle	11111	I
Halt	00100	H

TABLE 8.9   Starting Delimiter

Function or 4-bit group (4B)	Group code (5B)	Symbol
First symbol of sequential SD pair	11000	J
Second symbol of sequential SD pair	10001	K

network may contain a combination of 16 and 48-bit address entities (see Fig. 8.12).

Source address (SA; 16 or 48 MAC bits).   This field specifies the device that sent the frame (see Fig. 8.12).

Routing information (RI; 8 to 120 MAC bits).   The MAC-2 standard (revision 4.0 or greater) defines the optional RI field. (This field is not present in the original MAC standard X3.139-1987.) This field is used only by those networks that support source routing, and the field is present in those frames in which the routing information indicator (RII) bit in the SA symbol is set to 1. In MAC-2, the first bit of the 48-bit SA field functions as the RII (the first bit of a 16-bit SA field is currently reserved and set to 0). The RII is set to 1 to indicate the presence of the routing information field in this frame; otherwise the RII-bit is set to 0. The second bit is a universal/local (U/L) bit. The U/L bit is set to 0 to indicate use of an address plan that is universally administered by a central authority (such as IEEE) or set to 1 to indicate the use of a locally administered, user-defined address plan. If present, the RI field contains between 2 and 30 symbol pairs (octets). The first octet contains the length of the RI field. This length field has the format xxxNNNN0, where NNNN0 is the number of octets in the entire field (takes an even value between 2 and 30). The remaining 1 to 29 octets specify the path that the frame is required to take in order to get to its destination. This format conforms with the IEEE 802.1D standard (Chap. 5).

Information.   This field contains an LLC PDU or information related to a control operation. Information is encoded 4 bits at a time according to Table 8.10. For example, the bit sequence

$$0101/0001/1111/1010$$

is encoded as

$$5/1/F/A$$

namely

$$01011/01001/11101/10110$$

This will be described further in a later section.

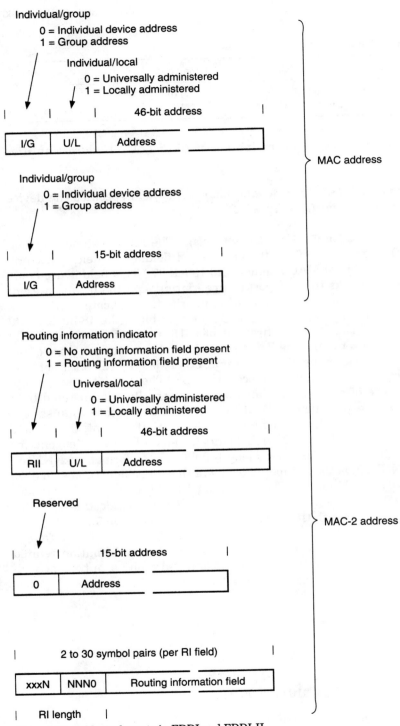

Figure 8.12 Address formats in FDDI and FDDI-II.

TABLE 8.10 Data Symbols

Function or 4-bit group (4B)	Group code (5B)	Symbol
0000	11110	0
0001	01001	1
0010	10100	2
0011	10101	3
0100	01010	4
0101	01011	5
0110	01110	6
0111	01111	7
1000	10010	8
1001	10011	9
1010	10110	A
1011	10111	B
1100	11010	C
1101	11011	D
1110	11100	E
1111	11101	F

Frame check sequence (FCS; 32 MAC bits). This field contains a 4-octet cyclic redundancy code covering the following four fields: FC, DA, DA, and information field.

Ending delimiter (ED; 4 or 8 MAC bits). This field flags the end of the frame (expert for the FS field which follows immediately). The delimiter is 8 bits for a token (two nondata T symbols) and 4 bits long (one T symbol) for all other frames. See Table 8.11.

Frame status (FS). This field contains status information. It has the form

Error detected (E)	Address recognized (A)	Frame copied (F)

Each indicator is represented by a symbol, which is R for "false" and S for "true." See Table 8.12. 4B/5B codes not shown in Tables 8.8 to 8.12 are not (re)transmitted because they violate consecutive code-bit

TABLE 8.11 Ending Delimiter

Function or 4-bit group (4B)	Group code (5B)	Symbol
Used to terminate data stream	01101	T

TABLE 8.12   Control Indicators

Function or 4-bit group (4B)	Group code (5B)	Symbol
Logical ZERO (reset)	00111	R
Logical ONE (set)	11001	S

zeros or duty cycle requirements. Some of the codes shown below are nonetheless interpreted as a Halt if somehow received:

00001	Void or Halt
00010	Void or Halt
00011	Void
00101	Void
00110	Void
01000	Void or Halt
01100	Void
10000	Void or Halt

8.3.2   FDDI MAC procedure

A device can send information only when it has ownership of the token. The device captures the token by absorbing the remainder of the token before the entire FC field passes by. After the captured token is completely received, the device begins transmitting frames, until it has no more frames to transmit or a time limit expires.

All other devices monitor the network and repeat, bit by bit (i.e., retransmit) each passing frame, as it comes by. In the earlier example of Fig. 8.10, this was shown by indicating that a given device "passes" the frame. Each device can check passing frames for errors and can set the E (error) indicator in the FS field if an error is detected. If a device recognizes its own address in the frame's DA field, it sets the A (address) indicator as the frame is retransmitted on the network; it may also copy the frame, setting the C (copy) indicator.

Eventually the frame will again reach the originator. The FS field (the E, A, and C indicators) allows the originator device, as it absorbs a frame that it previously transmitted, to identify three conditions:

■ Destination nonexistent or nonactive

■ Destination active but frame not copied

■ Frame copied

Note, however, if an error or failure to receive condition is discovered, the MAC protocol entity does not attempt to retransmit the frame, but reports the condition to LLC. It is the responsibility of LLC or some higher-layer protocol to initiate corrective action, if required.

The operation described above makes it clear why FDDI may not be able to support high sustained data rates to each device, as would be the case in LATM where each user is "guaranteed" 155 Mb/s. (This is discussed in the next chapter.) Besides having to wait for the token, each device on the network must repeat all data from all other devices. Note that an FDDI ring can support up to 500 directly attached devices. This implies that 500 devices must all be retransmitting data. (Note, however, that the transmission is immediately pipelined, so that the nodal delay is small.) In an ATM-based LAN (described in the next chapter), no device is involved in another device's transmission, and there are no tokens: each device obtains its own 155 Mb/s in a logical and physical star arrangement.

8.3.3 FDDI capacity allocation

FDDI supports two types of traffic: asynchronous and synchronous. Each device is allocated a portion of the total capacity. The synchronous allocation of a device represents the fraction of the time that the device is guaranteed for transmission during one rotation of the token. Capacity that is not allocated or that is allocated but not used, is available for the transmission of additional frames, referred to as asynchronous frames. A target token rotation time (TTRT) is defined. Device i is provided a synchronous allocation (SA_i), which may vary among devices and may also be zero. The allocations must be such that

$$D_{max} + F_{max} + \text{token time} + \sum_{i=1}^{n} SA_i \leq TTRT$$

where D_{max} = maximum propagation time, that is, propagation for one complete journey along the ring

F_{max} = time to transmit a maximum-length frame (4500 octets)

Asynchronous traffic can be subdivided into eight levels of priority. In addition, FDDI provides a mechanism to allow two devices to monopolize the asynchronous capacity for an extended exchange.[7]

Each station on the FDDI ring uses three timers to regulate its operation. Each device must have knowledge of the following parameters: TTRT, token-rotation timer, token-holding timer, and the valid transmission timer (see Table 8.13).

The responsibility for monitoring the proper operation of the token ring algorithm is distributed among all FDDI devices directly con-

TABLE 8.13 Device Counters

Token-rotation timer (TRT)	Used to clock the period between the receipt of tokens. The TRT is initialized to different values depending on the condition of the ring. During steady-state operation, the TRT expires when the TTRT has been exceeded. Devices negotiate the value for TTRT using the claim process.
Token-holding timer (THT)	Controls the length of time that a station can initiate asynchronous frames. A device holding the token can begin asynchronous transmissions if the token-holding timer has not expired. The THT is initialized with the value equal to the difference between the arrival of the token and the TTRT.
Valid transmission timer (VTT)	Measures the period between valid transmissions on the ring. It is used to detect excessive ring noise, token loss, and other faults. When the device receives a valid frame or token, the valid transmission timer resets. If the VTT expires, then the device starts a ring initialization sequence.

nected to the network. Each device monitors the ring for invalid conditions requiring ring initialization. Invalid conditions include an extended period of inactivity or incorrect activity. Each device maintains the TRT counter indicating how long it has been since it last noted a valid token. If this time exceeds *two times TTRT*, an error condition is assumed and the token is considered lost. Three processes are involved to clear these conditions:

- Claim token process
- Initialization process
- Beacon process

Any device detecting a lost token initiates the claim-token process by issuing a sequence of *claim* frames. The purpose of this process is to negotiate the value to be assigned to TTRT and to resolve contention among devices attempting to initialize the FDDI network. Each claiming device sends a continuous stream of claim frames. The information field of the claim frame contains the device's bid for the value TTRT. Each claiming device inspects incoming claim frames and either ceases to transmit its own claim frames and just repeats incoming frames or continues to transmit its own claim frames and absorbs incoming frames, according to a simple locally executable arbitration hierarchy. The process completes when one device receives its own claim frame, which has made a complete journey of the ring without being preempted. At this point, the FDDI network is permeated with that device's claim frames and all other devices have yielded. All devices store the value of TTRT contained in the latest received claim frame. The result is that the smallest requested value for TTRT is used by all devices to allocate capacity. The device that has won then issues a (nonrestricted) token. If the claim token process is inconclusive, a device may escalate to the beacon process.[7]

The beacon process is used to manage a serious ring failure such as a break in the ring. In this process, a device that suspects a ring failure transmits a continuous stream of beacon frames. A device always yields to a beacon frame received from an upstream device. Consequently, if the serious ring failure persists, the beacon frames of the device immediately downstream from the serious ring failure will normally be propagated. If a device in the beacon process receives its own beacon frames, it assumes that the ring has been restored, and it initiates the claim-token process.

8.3.4 FDDI PHY

The medium-independent portion of the physical layer specification (see Fig. 8.1) is called the physical layer protocol (PHY). The physical layer protocol corresponds to the upper sublayer of the OSI physical layer. It deals with such issues as the encoding scheme, clock synchronization (methods whereby a receiving device can use the incoming data to synchronize its clock with that of the transmitting device), timing jitter management, and data framing. PHY also provides a *smoothing* function to prevent frames from being lost and a *repeat filter* to prevent the propagation of code violations and invalid line states. Figure 8.13 depicts the logical function of the PHY module.

Bit encoding. FDDI uses intensity modulation to transmit information over a fiber. A binary one is represented by the presence of a pulse of energy (just below the visible domain, in the infrared region) of a certain duration; binary zero is represented by the absence of light for the same period. This scheme must recover synchronization from the signal stream; this is achieved by looking for (frequent) signal transitions (from energy to no energy and from no energy to energy). The problem is that a long string of 1s or 0s produces no transition. The solution to this problem is to encode the binary data to guarantee the presence of transition and then to present the encoded data to the optical source for transmission.

For MMF and SMF, FDDI uses a serial-based band transmission scheme based on the 4B/5B method already alluded to (twisted-pair systems may employ more sophisticated methods). In this scheme encoding

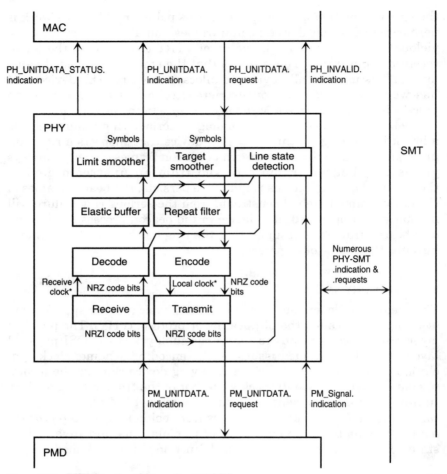

Figure 8.13 PHY functional view. * = 125 MHz.

is done 4 bits at a time, which must be buffered in order to determine which five-pulse symbol to transmit. This method is also called "4-out-of-5 encoding." For proper functioning of the FDDI network, PHY needs to keep all of the repeaters on the FDDI ring synchronized. In 4B/5B encoding, 5-pulse codes have been chosen in such a manner as to provide a guaranteed number of transitions to allow for robust clock recovery. Each repeater recovers clocking information from incoming signals by means of the transitions provided by the encoding scheme. As frames circulate around the ring, each repeater receives the data and recovers the clocking, enabling the receiver to maintain bit synchronization.

Figure 8.14 depicts an example graphically (refer to the earlier tables for actual values). In effect, each set of 4 bits is encoded as 5 bits. The efficiency is 80 percent: 100 Mb/s is achieved with 125

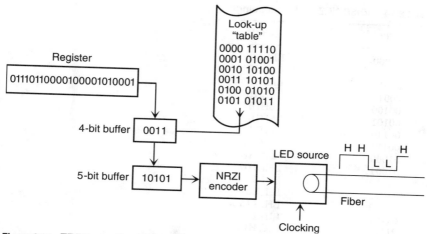

Figure 8.14   FDDI encoding (NRZI 4B/5B).

Mbaud (baud represents signaling elements per second). Note that only 16 combinations out of the actual 32 combinations possible with the 5-pulse code are required. The FDDI designers could then pick 16 "good" codes and declare the other patterns as either invalid or reserved for special control symbols. There are many ways in which 16 patterns can be chosen out of 32, the simplest one being the lexicographic method. The actual selection process was as follows: pick 5-pulse codes such that transitions are present at least twice.

Actually, a two-stage coding method is employed. After the 4B/5B code is derived from a nibble, giving rise to a nonreturn-to-zero (NRZ) code, the binary stream is encoded using a modified nonreturn-to-zero inverted (NRZI). With this scheme a binary 1 is represented with a transition at the beginning of the bit interval, and a binary 0 is represented with no transition at the beginning of the bit interval; there are no other transitions. The advantage of this scheme is that the signal is decoded by comparing the polarity of adjacent signal elements rather than the absolute value of a signal element, implying that it is generally more reliable to detect a transition in the presence of noise, attenuation, and distortion than to compare a value to a threshold. Table 8.14 lists for completeness the actual NRZI encoding used in FDDI.

As implied in Sec. 8.3.1, FDDI defines three types of symbols:

- *Data symbols* represent the actual data being sent
- *Line state symbols* show the connection status between neighbor devices
- *Control indicator symbols* show the status of the frame (devices set control symbols as the frame moves around the ring)

TABLE 8.14   4B/5B NRZI FDDI Encodings

NRZ	NRZI*
00000	LLLLL
00001	LLLLH†
00010	LLLHH†
00011	LLLHL†
00100	LLHHH
00101	LLHHL†
00110	LLHLL†
00111	LLHLH
01000	LHHHH†
01001	LHHHL
01010	LHHLL
01011	LHHLH
01100	LHLLL†
01101	LHLLH
01110	LHLHH
01111	LHLHL
10000	HHHHH†
10001	HHHHL
10010	HHHLL
10011	HHHLH
10100	HHLLL
10101	HHLLH
10110	HHLHH
10111	HHLHL
11000	HLLLL
11001	HLLLH
11010	HLLHH
11011	HLLHL
11100	HLHHH
11101	HLHHL
11110	HLHLL
11111	HLHLH

*Or, complement of the representation shown, depending on previous state.
†Violation or Halt.

Timing jitter.   The recovered clock may deviate from the timing of the transmitter because of signal impairments. This results in timing errors, known as timing jitter. As each device retransmits received data, it sends a signal with no distortion; however, the timing error is not eliminated. The cumulative effect of the jitter is to cause bit latency. Unless the latency of the ring remains constant, bits will be added as the latency increases and bits will be dropped (not retransmitted) as the latency of the ring decreases.

FDDI deals with timing jitter by using a distributed clocking mechanism with elastic buffers. Each repeater uses its own autonomous clock to transmit bits it receives from its MAC layer. For repeating incoming data, a buffer is interposed between the receiver and the transmitter. Data is clocked into the buffer at the clock rate recovered from the incoming stream, but it is clocked out of the buffer at the device's own clock rate.

The buffer has a nominal capacity of 10 bits and contracts or expands. At any time, the buffer contains a certain number of bits; as new bits come in, they are placed in the buffer. These bits experience a delay equal to the time it takes to transmit the bits ahead of it already in the buffer. If the received signal is faster than the repeater's clock, the buffer expands to avoid losing bits. If the received signal is slow, the buffer contracts to avoid adding bits to the repeated bit stream. The buffer in each repeater is initialized to its center position each time that it begins to receive a frame, during the preamble phase. This increases or decreases the length of the preamble, initially transmitted as 16 symbols, as it proceeds around the ring.

8.3.5   FDDI PMD

PMD defines how devices physically attach to the FDDI ring and how devices are physically interconnected on the network. PMD includes specifications for power levels, characteristics of the optical transmitter and receiver, acceptable bit-error rates, media type, connectors, and so on (see Table 8.15).

As shown in Fig. 8.1 a number of PMDs are supported by FDDI, including the currently evolving twisted-pair PMD. These include:

- MMF-PMD: multimode fiber
- SMF-PMD: single-mode fiber
- LCF-PMD: low-cost fiber
- TP-PMD: twisted-pair
- SPM: SONET physical layer mapping (an alternative to PMD)

The optical transmitter converts the encoded data from the PHY to a series of light pulses which are carried over the fiber optic cable. The optical receiver converts the received signals back to encoded

TABLE 8.15   PMD Functions (High-Level)

Optical transmitters and receivers
Fiber optic cable
Medium interface connector
Optical bypass relay (optional)

TABLE 8.16   PMD Parameters (Partial)

Wavelength of optical signal	(Nominal) optical wavelength refers to the approximate wavelength of optical signal used to carry information over the fiber cable. Light sources emit pulses at wavelengths such as 850 nm, 1300 nm, or 1550 nm. FDDI specifies 1300 nm.
Attenuation (amount of power loss) in the cable and other components (splices, connectors, etc.)	Measures the amount of power that is lost as the signal travels from the transmitter to the receiver. The attenuation of a link is determined by the attenuation per unit length (dB/km), which depends on the medium and, clearly, the length of the link. The MMF PMD standard specifies a maximum cable attenuation of 1.5 dB/km at 1300 nm.
Link-loss budget	Total amount of loss that can be introduced and still have the optical system work is called the link-loss budget. The link-loss budget is the difference between the minimum transmitter power and the receiver sensitivity minus any power penalty. The MMF PMD standard specifies a power budget of 11.0 dB.
Type of cable	The MMF PMD standard specifies 62.5/125 μm, multimode fiber optic cable. Single-mode fiber is currently nearing approval as another PMD standard. Also under active study is a copper STP/UTP PMD standard.

data. FDDI standards specify a light-emitting diode (LED) as the optical transmitter for multimode fiber optic cable. For single-mode fiber optic cable, laser diodes (LDs) are usually used as the optical transmitter. The PMD standard specifies size and optical characteristics for the fiber optic cable in the areas listed in Table 8.16.

MMF was the original PMD. The wavelength specified for data transmission is 1300 nm. The physical dimension specified in the standard is 62.5/125 μm (the first number represents the diameter of the core of the fiber; the second number the outer diameter of the cladding layer that surrounds the core). (The MMF standard also lists as alternatives 50/125, 82/125, and 100/140 μm). The connector is known as medium interface connector (MIC). The MIC properly aligns the fiber with the transmit-receive optics in the node. The connector consists of a keyed plug and a keyed receptacle. Keying ensures that the plug is installed correctly. (ST-type connectors are also commonly used as a lower-cost alternative.)[3]

An increasing number of FDDI users do not want to incur the expense of purchasing, installing, and managing their own optical fiber transmission facilities, since this entails specialized skills and test equipment. Instead, they prefer to use high-fiber transmission services that are available from carriers. Note that in this context, the word *service* is critical, since a carrier has no incentive in dedicating one pair (or even two) of fiber which could carry 2 Gb/s to support only 100 Mb/s. Such an approach would eliminate the barriers to FDDI extension and expansion because of right-of-way issues. Hence, the users need to interface FDDI seamlessly over a shared SONET-based infrastructure. In 1989, X3T9.5 started to develop an interface between FDDI's PHY protocol and the SONET standards. While a mapping has been established, no X3 document is available as of time of publication.

8.3.6 Station management

FDDI devices can have multiple instances of PMD, PHY, and MAC entities, but only one SMT entity. SMT covers three major functions: connection management (CMT), ring management (RMT), and SMT frame services (see Fig. 8.15).

Connection management.   CMT is concerned with the insertion of devices onto the FDDI network and the removal of devices from the FDDI network. This involves establishing or terminating a physical link between adjacent ports and the connection of ports to MAC entities. Since FDDI devices can have multiple occurrences of PHYs and MACs, one of the functions of CMT is to manage the configuration switch that connects PHYs to MACs and to other PHYs within a node. Connection management functions include:

- Connecting nodes to the FDDI network
- Connecting PHYs in separate nodes (including concentrators)
- Using trace diagnostics to identify and isolate a faulty component

To support these functions, CMT encompasses the following submodules (see Fig. 8.15):

- Entity coordination management (ECM)
- Physical connection management (PCM)
- Configuration management (CM)

ECM deals with the medium interface to the FDDI ring, including port coordination and operation of the (optional) optical bypass switch associated with that device. PCM deals with the point-to-point physical links between adjacent PHY-PMD pairs, including initializing the link and testing the quality of the link (called "link confidence"). CM deals with configuring PHY and MAC entities within a node; it is concerned with the internal organization of the device entities.

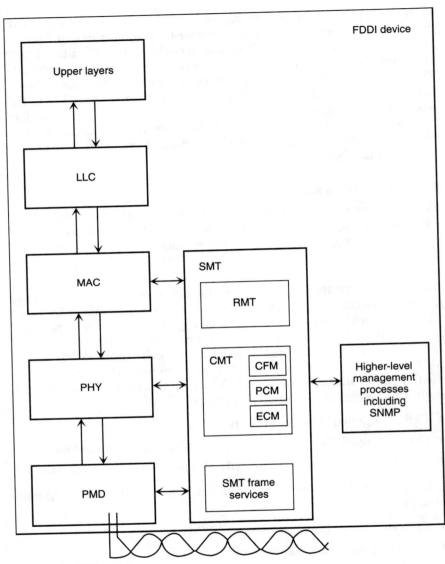

Figure 8.15 SMT modules.

Initializing is accomplished by signaling between the adjacent ports. One port transmits a continuous stream of symbols until the neighbor responds with another stream of symbols. PCM repeats a number of these request-response exchanges to communicate the information shown in Table 8.17. Once the connection has been verified, configuration management is invoked.

TABLE 8.17 Information Sent at Initialization

Port type (A, B, M, S)

Willingness to establish a link

Duration of the link confidence test performed

Availability of the MAC entity for a link confidence test

Outcome of the link confidence test

Availability of the MAC for a local loop test

Intent to place a MAC in the connection if established

Ring management. RMT receives status information from MAC and CM. RMT then reports this status to station management and higher-level processes. Functions supported by RMT include:

- Stuck beacon detection

- Resolution of problems through the trace process

- Detection of duplicate addresses

A stuck beacon indicates that a device is locked into sending continuous beacon frames. A device using beacons should eventually receive either a beacon from an upstream device or its own beacon. If neither event occurs, the device will transmit its own beacon indefinitely. A *stuck-beacon timer* controlled by RMT measures the duration of beacon transmission. If a time limit is exceeded, RMT initiates a stuck-beacon recovery procedure. The procedure entails the transmission of a directed beacon with a multicast address to all directly connected devices, informing them of the stuck-beacon condition. The directed beacons are sent for a sufficiently long time to assure that they are seen by all the directly connected devices. If the stuck-beacon condition is still unresolved, a trace function is initiated.

The trace function uses PHY signaling to recover from a stuck-beacon condition. The goal is to localize the fault to the beaconing device and its nearest upstream neighbor.

If two or more devices (more precisely, MAC entities) erroneously have the same address, the network clearly cannot function properly. Duplicate address detection is performed during ring initialization. If two or more devices have the same address, at least one of the devices will experience one of the conditions listed in Table 8.18.

When a device detects the duplicate-address condition, it can respond by changing its MAC address, configuring the MAC to lose the claim process and disabling its LLC services, or removing itself from the ring.

TABLE 8.18   Symptoms of Duplicate Address

Receive its own beacon while issuing claim frames for longer than D_{max}	Indicates that the other duplicate is sending beacon frames while this duplicate is sending claim frames
Receive its own claim frame while issuing beacon frames for longer than D_{max}	Indicates that the other duplicate is sending claim frames while this is sending beacon frames
Receive its own claim frame for a period of time greater than D_{max} after having "won" the claim-token contest	Indicates that the other duplicate is sending claim frames while this duplicate has stopped claiming and has issued a token
Receive its own claim frame with different value of TTRT	Indicates that duplicate devices with different requested TTRT values are both claiming

SMT frame services.   The frame services portion of SMT deals with the management of the device after the ring has achieved an operational state. Such services are implemented by a set of SMT frames, shown in Table 8.19. These services are implemented by different SMT *frame classes* and *types*. Frame class identifies the function that the frame performs, such as neighborhood information frame (NIF) and station information frame (SIF). Frame type designates whether the frame is an announcement, a request, or a response to a request.

TABLE 8.19   SMT Frames

Neighborhood information frame	Used to transmit own address and basic device descriptor to downstream neighbors. Each device periodically issues the frame using "next station addressing" (a special addressing mode that permits a device to transmit a frame to the next device in the path without the address of that device).
Station information frame	Used to request and supply a device's configuration and operating information
Echo frame	Used for SMT-to-SMT loop-back testing
Resource allocation frame	Used to support network policies for the allocation of resources (e.g., the allocation of synchronous bandwidth to the devices)
Request-denied frame	Used in response to an unsupported optional frame class or unsupported version ID
Status-report frame	Used by devices to periodically announce device status
Parameter management frame	Used for remote management of device attributes via the parameter management protocol
Extended-services frame	Used for user-defined frames that extend SMT services

The two most important frames are:[3]

Neighborhood information frames. Devices use neighborhood information frames (NIF) to periodically announce their addresses to downstream neighbors. Each device in the ring makes such an announcement approximately every 30 s by sending an NIF that uses next station addressing (NSA). NSA, a special addressing mode, permits a device to send a frame to the next device in the token path without knowing the address of that device. After 30 s, each device in the ring knows the address of its upstream neighbor. These addresses can be used to create a logical ring map showing the order in which each device appears in the token path.

Station information frames. Devices use station information frames to exchange more detailed information about their characteristics and configuration. The configuration information is used to create a logical ring map in much the same way as that created from NIFs. In addition, SIFs contain information about the attachment status of each port in a concentrator. This information can be used to create a physical ring map that shows the position of each device not only in the token path but in the topology as well.

8.3.7 Comparison with 802.3 and 802.5

Table 8.20 provides a synopsis of the key technical differences between FDDI and Ethernet and token ring LANs.

8.4 FDDI-II

FDDI-II is an upward-compatible extension to FDDI that adds the ability to support circuit-switched traffic, in addition to the packet-mode traffic supported by the original FDDI.[6] FDDI is not suitable for maintaining a continuous, constant-data-rate connection between two devices, for applications such as voice, video, and multimedia. Even the synchronous traffic class of FDDI only guarantees a minimum sustained data rate; it does not provide a uniform information stream with no interframe variability. Such a continuous, constant data stream is typical of circuit-switched applications, such as digitized voice or video.[8] Compressed digital video based on the motion picture expert group (MPEG) compression standards (MPEG-1 or MPEG-2) can be supported; video based on CCITT H.200 standards or digital video interactive (DVI) encoding can also be supported.[9]

FDDI-II provides a circuit-switched service while maintaining the token-controlled packet-switched service of the original FDDI. Hence, with FDDI-II systems it is possible to set up and maintain a constant-

TABLE 8.20   Top-Level Comparison between FDDI and IEEE LANs

Category	FDDI-I	IEEE 802.5 token ring	IEEE 802.3 Ethernet
Physical topology	Ring, star, hierarchical star	Ring, star	Ring, star, hierarchical star
Logical topology	Dual ring, dual ring of trees	Ring	Bus
Media	Single-mode fiber,* multimode fiber, twisted-pair,* SONET*	Twisted-pair, fiber*	Various coaxials, twisted-pair (10Base-T), fiber (FOIRL†) wireless*
Bandwidth (Mb/s)	100	4 or 16*‡	10‡
Media access	TTP	Token passing	CSMA/CD
Token acquisition	Absorption	By setting a status bit, converts token into a frame	na
Token release	After transmit or time out	After receive (4 Mb/s) or after transmit (16 Mb/s)	na
Maximum frame size (octets)	4500	4500 (4 Mb/s), 18,000 (16 Mb/s)	1518
Number of nodes	500	260	1024
Distance between nodes	2 km (1.2 mi) for MMF	300 m (1000 ft) station to hub (4 Mb/s); in practice, 100 m (330 ft) is recommended for both systems	2.8 km (1.7 mi)
Maximum network span	100 km (60 mi)	Depends on configuration	2.8 km (1.7 mi)

*Not standardized/not yet a standard.
†Fiber optic inter-repeater link.
‡Other options allowed in standards.

rate connection between two devices. Instead of using embedded addresses, as was the case with FDDI's frames, the connection is established on the basis of a prior agreement, which may have been negotiated using packet messages by some other suitable mechanism. However, one of the limitations of FDDI-II is that the total bandwidth remains 100 Mb/s, so the averaged per capita bandwidth is still relatively low, as seen in Table 8.21, which also juxtaposed the comparable telecommunications service.

The technique used in FDDI-II for providing circuit-switched service is to impose a 125-μs frame structure on the FDDI network. A

TABLE 8.21   Maximum Theoretical Bandwidth per User in FDDI-II

Active users	% utilization (no. users/500)*	Averaged per user bandwidth	Telecom equivalent
500	100	200 kb/s	Basic-rate narrowband ISDN
250	50	400 kb/s	H0 narrowband ISDN
125	25	800 kb/s	Fractional T1
62	12.5	1.6 Mb/s	H11 narrowband ISDN
31	6.25	3.3 Mb/s	DS1C
15	3.1	6.6 Mb/s	DS2
2	0.4	50 Mb/s	DS3/SONET
1	0.2	100 Mb/s	Just shy of B-ISDN

*Compared to the maximum possible number of users that can access the system, namely, 500.

circuit-switched connection consists of regularly repeating time slots in the frame. (This mode of transmission is also known as isochronous.)

In an FDDI-II device, the physical layer and the station management are the same as for the original FDDI (see Fig. 8.1). At the MAC level, two new components, referred to as hybrid ring control, are added: the hybrid multiplexer (HMUX) and an isochronous MAC. The IMAC module provides the interface between FDDI and the isochronous service, represented by the circuit-switched multiplexer (CS-MUX). The HMUX multiplexes the packet data from the MAC and the isochronous data from IMAC.

An FDDI-II network can operate in either *basic* or *hybrid* mode. In basic mode, only the packet-switched service, controlled by a token discipline, is available. In this mode, the network operates as in the original FDDI. In hybrid mode, both packet and circuit services are available. An FDDI-II network typically starts operation in the basic mode to establish and propagate the timers and parameters necessary for the TTP, then switches to hybrid mode.

Hybrid mode FDDI-II employs a continuously repeating PDU referred to as a cycle. The cycle is a framing structure similar in principle to that used in synchronous transmission systems (e.g., SONET; see Chap. 9). The contents of the cycle are visible to all devices as it circulates around the ring. A device called the cycle master generates a new cycle 8000 times per s, or once every 125 µs. At 100 Mb/s the cycle size is 12,500 bits. As each cycle completes its journey around the ring, it is absorbed (stripped) by the cycle master.

Figure 8.16 depicts the format of the cycle. The cycle consists of the following elements:

- *Preamble.*   This is a five-symbol nondata stream. The actual size of the preamble may vary to maintain synchronization in the presence of jitter.

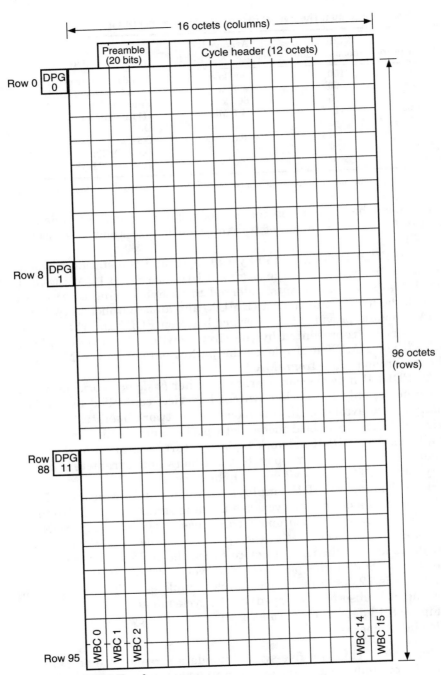

Figure 8.16   FDDI-II cycle.

TABLE 8.22 **Content of FDDI-II Header**

Starting delimiter	Used to indicate the beginning of a cycle; it consists of the JK symbol pair.
Synchronization control	Used to establish the synchronization state of the FDDI-II ring. A value of R indicates that synchronization has not yet been established and that the cycle may be legally interrupted by another cycle. The field is set to the symbol R (00111, logical 0, false) during hybrid mode initialization or by any device that detects loss of cycle synchronization by not receiving a cycle within 125 μs of the previous cycle. A value of S (11001, logical 1, true) indicates that synchronization has been established; this value can only be set by the cycle master.
Sequence control	Used to indicate the status of cycle sequencing. A value of R indicates that either the cycle sequence has not yet been established or that a cycle error has been detected. A value of S indicates that valid cycle sequence is established and devices can latch each cycle sequence value to compare to the cycle sequence value in the next cycle.
Cycle sequence (CS)	Takes the form NN where N is a data symbol. If the C_1 and C_2 fields both contain R, then the CS field is interpreted as containing a monitor rank. The monitor rank is used during the monitor contention process.
Programming template	Consists of 16 symbols, one for each WBC. An R value indicates that the corresponding WBC is part of the packet data channel, while an S indicates that the corresponding WBC is dedicated to isochronous traffic. The programming template is ready by all devices but may only be modified by the cycle master.
Isochronous maintenance channel	Used to carry isochronous traffic for maintenance purposes (use is outside the scope of the FDDI-II standard).

- *Cycle header.* This is a 12-octet header which contains information that defines the usage of the remainder of the cycle (see Table 8.22, based partially on Stallings[8]).

- *Dedicated packet group (DPG).* These 12 octets are always available for token-controlled packet transfer.

- *Wideband channels (WBC).* Each of 15 wideband channels consists of 96 octets per cycle.

Each of the wideband channels, at 96 octets, provides a capacity of 6.144 Mb/s. Each channel may be used for circuit switching or packet

TABLE 8.23   Capacity Allocation in FDDI-II

Category	Bits per cycle	Rate (Mb/s)
Overhead (cycle header plus preamble)	116	0.928
N channels of circuit-switched information	$N \times 768, N \le 16$	$N \times 6.144$
Packet data channel	$96 + (16 - N) \times 768$	$0.768 + (16 - N) \times 6.144$
Total	12,500	100

switching. If the channel is used for packet switching, then it is merged with the dedicated packet-group octets and with other WBCs set aside for packet switching, to form one large channel dedicated to packet switching. This channel, the packet data channel, is controlled by a token discipline. The capacity of the packet data channel is allocated on this channel using the FDDI-I MAC protocol. The minimum capacity of the packet data channel is 768 kb/s, and it can grow in increments of 6.144 Mb/s, to a maximum of 99.072 Mb/s.

The IMAC sublayer within the HRC controls the WBCs that are used for circuit-switched traffic. Each of fifteen 6.144-Mb/s wideband channels can support a single isochronous channel or may be subdivided by IMAC into a number of subchannels supporting simultaneous, independent, isochronous sessions between different pairs of FDDI-II devices, at a lower data rate. During normal operation, the activity on an FDDI-II network consists of a sequence of cycles generated by the cycle master. Devices communicate using circuit switching by sharing the use of a dedicated isochronous channel. Devices communicate using packet switching over the packet data channel, observing the rules imposed by the TTP. The capacity allocation of FDDI-II is accounted for as shown in Table 8.23.

Normally, the ring is configured to initialize in basic mode. Once basic mode is established and operating, one or more devices may attempt to move the network to hybrid mode by issuing a cycle. One monitor device can be preassigned this task, or all monitor devices may compete. Fairly complex procedures are required to achieve bandwidth allocation.

8.5   Efforts to Bring the Cost of FDDI to the Desktop Down

8.5.1   Processorless FDDI adapter design*

As discussed above, SMT defines protocols and services for LAN layer management. Each FDDI station contains one SMT entity. The SMT

---

*This section was provided by Gautam Chanda of 3Com Corporation.

initializes the network, monitors error rates and fault conditions in each network segment, and automatically reconfigures the network to isolate problem links. SMT components are CMT, which includes PCM, and RMT. CMT manages the physical connections (links) between adjacent FDDI nodes, initializes adjacent stations' PHYs, and manages signaling. RMT performs duplicate address conditions that prevent the ring from becoming operational, and it also manages the logical rings that comprise an FDDI network.

When designing an FDDI adapter, there are two possible design and architectural alternatives: to have an on-board processor to execute all the SMT and any other required management functions or to have no on-board processor and thereby off-load all the SMT and other software execution to the host processor. By careful analysis of different issues throughout the FDDI adapter design cycle, one can arrive at an adapter design that offers high performance, yet be flexible and cost-effective.

With a processorless design, typically the overall adapter design is quite straightforward and lends itself to a short development cycle, resulting in early market entry and low cost to the end user. This is because the processorless adapter architecture is a simpler design that includes little additional hardware logic besides the FDDI chipset. However, the processorless adapter needs some special design considerations, most importantly the issues related to the large bulk of SMT software to be executed by the host processor.

SMT typically has few real-time requirements except for PCM, which manages the links between the adjacent FDDI nodes. This real-time requirement is the 3 ms PC_REACT timing in PCM. Therefore, in a processorless adapter design the host processor will have to be responsible for this 3-ms interrupt latency. Depending on the host processor load for other processing at any given time, this 3-ms interrupt latency may not be guaranteed. Some FDDI chipsets in the market already provide the basic PCM functions in their PHY chip; therefore, they are more suitable for the processorless adapter design.

The SMT software is typically about 200K of code and data. In a processorless design, the host processor memory space will be consumed by this additional software executed by the host. In memory-constrained environments, such as DOS, this can be a problem. In such operating system environments, some software tricks, such as moving the SMT software into extended memory, may have to be implemented to alleviate these problems.

With on-board processor architecture, all the SMT software as well as any associated network management software is executed by the on-board processor and thereby isolate the host processor from any SMT-related tasks. On-board processor design is typically a more challenging task owing to more on-board logic and the more constrained on-board software debugging environment. This will likely result in a more expensive design and longer development cycle, so it will be longer in coming to market. On the positive side, the on-board processor design is a more elegant architecture and, overall, a much cleaner design than the processorless design.

One major design consideration for the adapter is how to maximize the system bus utilization to keep up with the network traffic, which theoretically can be up to 100 Mbps. For high-speed system buses such as Sbus or VME, where, even with bus contentions, typical bus throughput will exceed the peak FDDI traffic rate, this may not be a major design consideration. For slower-system buses, such as EISA, the adapter design has to consider ways to match the data rates between the system bus and the FDDI network so as to minimize losing packets and degrading the overall performance severely.

8.5.2   FDDI twisted-pair efforts*

In examining the costs associated with FDDI, it quickly becomes evident that the single largest cost contribution factor lies with the choice of optical links. There are two major reasons for this. First, fiber optic cable plants are not widely deployed and are relatively expensive to install.[10] Most office buildings and factories have a large installed base of LANs which use unshielded twisted-pair (and some shielded twisted-pair), which company managers are reluctant to scrap. The second reason is that optical transceivers that can meet the FDDI-11 dB power budget requirements are fairly expensive because of relatively low yields in the manufacturing process.

Several groups of companies have proposed different alternative PMD standards within the ANSI X3T9.5 committee. The twisted-pair PMD is a response to the user concern over the cost of fiber installation and the high cost of the fiber transceivers. Most users have existing cable plants that use voice-grade unshielded twisted-pair (UTP), as well as data-grade unshielded twisted-pair (DTP), particularly in newer installations for 10Base-T Ethernet LANs. According to surveys, at most installations the distance between the wiring closets to the desktop is at least 100 m, if not more. Therefore, any ANSI standard for twisted-pair has to work for at least 100 m. In the recent past, three different twisted-pair PMD alternatives have been proposed in the ANSI committee and an intense discussion over preferred cabling and PMD standard has followed in the ANSI committee.

The initial twisted-pair proposal on the table in the ANSI committee was for shielded twisted-pair (IBM type 1 cable) from IBM, SynOptics, DEC, Chipcom, and Motorola. This alternative is already

---

*Portions of this section were provided by Gautam Chanda, 3Com Corporation.

demonstrated to be working for distances up to 100 m. The shielded twisted-pair (STP) PMD proposal will probably be more applicable to existing IBM installations that already use IBM type 1 cable.

Later, an industry consortium called Unshielded Twisted-Pair Development Forum (UDF) proposed an FDDI cabling standard that incorporates voice-grade UTP. It is still unclear if this proposed scheme will allow FDDI signals to travel 100 m and the emissions will be low enough to meet FCC regulations. In light of the difficulties in achieving 100 m distance for FDDI signals over voice-grade UTP cables, recently the ANSI committee voted to immediately move forward with a single FDDI PMD standard only for DTP and STP. The standard for voice-grade UTP was excluded because the committee felt that to have a single standard which covers voice-grade UTP, DTP, and STP will take more investigation and will delay the standard completion.

In June 1992, after considerable discussion over the previous two years, X3T9.5 endorsed a method proposed by Crescendo Communications Inc.: multiline transmission 3 (MLT-3) for unshielded twisted-pair transmission. Final acceptance of the standard is expected by 1993. Twisted-pair FDDI has the chance of being more cost-effective compared to MMF and SMF systems. Equipment supporting this technology, also known as copper distributed data interface (CDDI), is starting to appear. At an entry-level price of $995, these cards are price-compatible with high-end Ethernet and token ring adapters. Yet, some users still feel the overall price is too high; concentrator prices in the $500 to $1000 range per port are being sought by these users.[11] (For comparison, note that 155 Mb/s copper links to the desktop using SONET and ATM, discussed in the next chapter, are also becoming feasible).

The second developing standard results from the fact that some optical-component manufacturers are asking a lower price for components that have slightly lower performance than required to meet the required 11-dB power budget. Given the previous observation about horizontal distances, a strong case in the standard committee was made for a PMD that is optically compatible with the standard multimode PMD but uses a smaller power budget. This new PMD is the LCF PMD, and, as stated above, is in the early stages of standards development within the ANSI committee.

With this discussion on emerging alternative PMD standards as background, it is clear that users are somewhat confused over appropriate cabling and PMD for FDDI systems they may want to install now and in the near future. Therefore, adapter designs have to offer flexibility to change media supported by the adapter to accommodate new standards for STP, UTP, and low-cost fibers. Preferably, it should be possible to field-upgrade the adapters to new media of choice. Some of the better adapters meet these goals.

TABLE 8.24   FDDI Market ($M)

Year	World market	U.S. market	Cost of FDDI connection
1991	$750	30	$7000–$12,000
1995	$2000	500	$1000–$2000

8.6 Commercial Aspects

About 75 vendors now manufacture FDDI-based products, including 3Com, AMP, DEC, Apollo/HP, Apple, AT&T, Cabletron, Cisco, Coral, Fibermux, Fibronics, IBM, Motorola, Proteon, Sun Microsystems, Timeplex, and Wellfleet. Table 8.24 shows the U.S. and worldwide market for FDDI products (see Minoli[12] for sources of data).

Some studies suggest that FDDI sales will flatten out in 1994, owing to the pressure from ATM-based local networks (Chap. 9). At time of publication major vendor support for FDDI-II appeared in doubt, owing to the emergence of ATM-based alternatives. Only two vendors had prototype products using partly proprietary elements: AWA Defense Industries and MultiMedia LANs Inc. Only one chipset (National Semiconductor's) supported FDDI-II protocols.[13]

References

1. D. Minoli, *Telecommunications Technology Handbook,* Artech House, Norwood, Mass., 1991.
2. D. Minoli, "Fiber Distributed Data Interface Standards," Report MT20-340-101, Datapro, August 1988.
3. *A Primer to FDDI,* EC-H0750-42 LKG, Digital Equipment Corporation, 1991.
4. F. E. Ross and R. L. Fink, "Overview of FFOL: FDDI Follow-On LAN," *Computer Communications,* Jan./Feb. 1992, pp. 5 ff.
5. F. E. Ross and R. L. Fink, "Following the Fiber Distributed Interface," *IEEE Network,* March 1992, pp. 50 ff.
6. G. C. Kessler and D. A. Train, *Metropolitan Area Networks,* McGraw-Hill, New York, 1992.
7. W. Stallings, "FDDI, Part 1: MAC Protocol and Station Management," *Open Systems Communication,* April 1992, p. 1.
8. W. Stallings, "FDDI, Part 2: Physical Layer Specification and FDDI-II," *Open Systems Communication,* May 1992, p. 1.
9. A. Shah, "Multimedia Works Well over Existing Networks with Careful Design," *Computer Technology Review,* Fall 1992, pp. 33 ff.
10. D. Pryce, "Opposing Groups Struggle to Define Standards for FDDI Using Copper," *EDN,* vol. 37, no. 5, 1992, pp. 57 ff.
11. M. Dortch, "Microsoft Signs On as CDDI User," *Communications Week,* November 9, 1992, p. 16.
12. D. Minoli, *Enterprise Networking, Fractional T1 to SONET, Frame Relay to B-ISDN,* Artech House, Norwood, Mass., 1993.
13. M. Fahey, "Support Wavers for FDDI-II," *Lightwave,* October 1992, p. 1.

9

Third-Generation LANs: Gigabit Systems

9.1 Introduction

Efforts are now under way to develop customer premises networks (namely, LANs) that support very high speeds and at the same time are compatible with evolving public network communication architectures such as B-ISDN and cell relay. This eliminates (or minimizes) the cost of ancillary equipment such as bridges, routers, and gateways to provide an n-layer relaying function (in OSIRM terminology) between the private network and the public network. This section provides a state-of-the art view of these proposals. The discussion below is based on the network-compatible local ATM (LATM) specification of April 1992. Some of the concepts are subject to change but are presented here to provide a sense of the direction that third-generation LANs are taking. Work is currently being done under the auspices of the ATM Forum, which now consists of over 275 companies.[1–3] Many leading vendors are now pushing ATM instead of FDDI to provide high bandwidth to the desktop.[4]

As discussed in Chap. 2, first-generation LANs emerged in the late 1970s. They supported 10 Mb/s on metallic media and were employed mostly to provide local data transmission for business applications such as e-mail and word processing. Second-generation LANs, in the form of FDDI-based systems, emerged in the late 1980s. They supported 100 Mb/s on fiber media (with some possible extensions to metallic media) and have been employed mostly to support campus-based LAN backbone interconnection and data applications. Although FDDI-II can support isochronous traffic (voice and compressed digital video), its total throughput has remained in the 100 Mb/s range; additionally, there has been little movement in terms of products or deployment.

ATM (cell relay) is a high-bandwidth, low-delay switching and multiplexing technology. ATM is the technology of choice for evolving B-ISDN public networks, which are beginning to appear (and will see increased deployment in the near future), but ATM will also penetrate next-generation LANs at the core, as a premises technology.[5–8] These LANs are now under development to support a variety of new applications such as desk-to-desk video conferencing, multimedia conferencing, multimedia messaging, distance learning, imaging tasks (including computer-aided design and manufacturing), animation, data fusion, cooperative work (for example, joint-document editing), and supercomputer access.[9–13]

Multimedia applications now emerging may entail both desk-to-desk multimedia conferencing (between both local and remote users) and client-server applications, where the images, video libraries, and multimedia e-mail messages are stored in a server. Multipoint-to-multipoint connectivity is being sought. Other emerging data-intensive applications were discussed in Chap. 2 and elsewhere in this text.

Desk-to-desk videoconferencing might be done at 384 kb/s or even 1.544 Mb/s. A network supporting 20 to 30 such users (a typical work group and LAN configuration), would need a *sustained* throughput of 10 Mb/s to 45 Mb/s; additionally, these isochronous applications are delay-sensitive and so may not be able to make optimal use of a contention (Ethernet) or token-based data transfer discipline. Some high-quality graphics applications may require as much as 45 Mb/s per user. Should 20 or 30 users need simultaneous access to a server (for example, a group of radiologists reviewing x-rays), the sustained throughput would have to be 1.2 Gb/s. These rates are in line with the B-ISDN and cell relay rates of 155 Mb/s—also known as synchronous transport signal level 3 concatenated (STS-3c), 622 Mb/s (STS-12c), and 1.244 Gb/s (STS-24c). (This latter rate is not currently envisioned for B-ISDN.)

There are four goals for third-generation LANs.[1]

1. Provide real-time transport capabilities necessary for multimedia applications (particularly for video signals).

2. Provide scalable throughput that can be grown both on a per-device basis and on an aggregate basis. Per-device scalability allows a few devices to receive more bandwidth if their applications warrant it. Aggregate scalability allows migration in terms of attached devices.

3. Facilitate interworking between LANs and WANs. Currently,

routers and bridges (whether using T1 technology, frame relay, or SMDS) need to undertake relatively major protocol conversion in order to provide access to WAN services. These differences in LAN and WAN technology have slowed the extension of truly distributed LAN-based computing to environments beyond a campus.

4. Use, to the extent possible, evolving ATM/B-ISDN standards. This has the advantage of expediting development and deployment of the technology, since a set of about two dozen ATM/B-ISDN standards is already available. It also keeps the cost down, since chips developed for SONET, B-ISDN ATM, B-ISDN adaptation layer, and so on, can be utilized.

Three "user plane" protocol layers are needed to undertake communication:

1. The ATM layer (equating approximately to the upper part of the MAC layer), which has been found to meet the stated objectives of throughput, scalability, interworking, and consistency with international standards. The function of the ATM layer is to provide efficient multiplexing and switching, using cell relay mechanisms.

2. A layer below the ATM layer, corresponding to the physical layer. The function of the physical layer is to manage the actual medium-dependent transmission. Two mechanisms are under consideration: SONET framing and fibre channel standard (FCS) framing. SONET is the preferred approach.

3. A layer above the ATM layer, called the ATM adaptation layer (AAL). The function of the AAL is to insulate the upper layers of the LAN application protocols (e.g., TCP/IP) from the details of the ATM mechanism. A very simple adaptation layer has been designed to support efficient cellularization and integrate smoothly into existing upper-layer protocols. This AAL is known as AAL type 5 or simple and efficient adaptation layer (SEAL).[14]

In addition, a mechanism is needed in the "signaling plane," in order to manage connection between users. In a departure from connectionless IEEE-based MAC communication, initial directions in ATM-based LANs are to use connection-oriented communication. This would make the LAN consistent with current B-ISDN services, although a connectionless service is also under development for B-ISDN. Connection-oriented communication requires a call-setup phase; signaling is needed to accomplish this. Connections can be made on a PVC-basis (particularly for early 1993–1994 products, called Phase I) and on an SVC-basis (1995 and beyond products, called Phase II).

This chapter starts out with an ATM/B-ISDN primer. Local ATM principles are discussed next.

9.2   ATM/B-ISDN Primer

ATM is now rapidly being put into place to allow high-speed seamless interconnection of LANs and WANs.[5-8] The term *cell relay* is also used since ATM transports user cells (data units of 53 octets) reliably and expeditiously across the network interface to the destination (5 of the 53 octets are for "overhead" and 48 are for user information). A network supporting this protocol is said to support cell relay service. Cell relay is one of two "fast-packet" technologies that have entered the scene recently (the other being frame relay, already discussed in Chap. 7). Nearly all the required technical specifications for ATM have already been published; this started with a basic set in 1988, continued with a larger set in late 1990 and in June 1992 progressed to a still further enlarged set. A variety of vendors is now readying end-user products for 1993 market introduction; platform products have already been on the market for a year or so.

The section which follows addresses the motivation and use of the technology in the corporate environment. The next section provides the basic technical machinery to understand what ATM is and to enable the manager to make intelligent decisions with regard to its merits and appropriateness.[5,6]

9.2.1   ATM: Where can it be used?

ATM is a high-speed switching and transport technology that can be used in a local area environment, a wide area environment, or both. Its functionality corresponds to the physical layer and part of the data link layer of the OSIRM. ATM supports speeds of 155 Mb/s and 622 Mb/s now and can go as high as 10 Gb/s in the future. As an option, ATM will operate at DS3 (45 Mb/s); some are also looking at operating at DS1 (1.544 Mb/s).

In most implementations, ATM entails

- User devices which can segment PDUs generated by the user's upper layers (at or above network layer or more precisely at or above the "upper part" of the user's data link layer) into cells

- An interconnection apparatus

The interconnection apparatus can be as simple as a dedicated medium (local twisted-pair or fiber or a high-speed digital telecommunications circuit), or can consist of a switch or cluster of switches. The switches can be customer-owned, and can be simple wiring hubs supporting the ATM protocol; or they can be more complex fast-packet switches. Alternatively, the switches can be owned by a carrier. The user devices can be a workstation configured with an ATM card, a router configured for ATM-cell relay service, or another device such as a host or a video codec.

Figure 9.1 depicts a number of ATM-cell relay configurations which a LAN manager might have to support in the next year or two. Figure 9.1a shows two ATM-configured devices connected over a simple point-to-point local channel. Only two devices participate in this communication. Its function is mostly to allow the corporation to "play with the technology." Figure 9.1b depicts the same introductory usage of the technology over a dedicated point-to-point telecommunications channel. This approach is similar to the one taken by many early frame relay implementations. Only two routers participate in this communication.

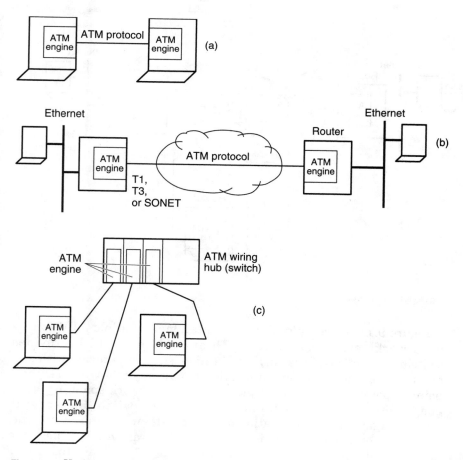

Figure 9.1 Various uses of LAN/WAN ATM.

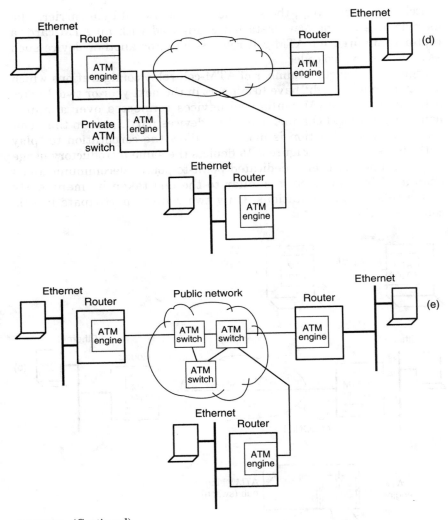

Figure 9.1 (*Continued*)

Figure 9.1c depicts a local implementation of multiple ATM-config-ured workstations connected with an ATM hub which provides switching functions; here, any workstation can communicate with any other workstation (the communication is now any-to-any rather than point-to-point). Figure 9.1d depicts a private network spanning some geographic distance with one private ATM switch and several routers. Again this configuration affords any-to-any connectivity. The advantage of this configuration is that if there are n sites, one only needs n links to the switch rather than $n(n-1)/2$ dedicated links, which would be required without the ATM switch.

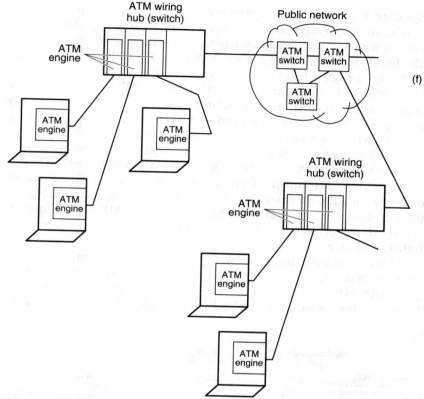

Figure 9.1 (*Continued*)

Figure 9.1e shows the situation where the cell relay service is provided in the public (WAN) network. (ATM switches are deployed in the public network.) This configuration relieves the manager from having to acquire and maintain the switch and the associated (long) communication lines, while at the same time providing reliable low-latency high-capacity switched connectivity across the network. Figure 9.1f depicts the most sophisticated and synergistic use of this technology at both the local area and wide area network. This is the goal sought by LATM.[1,2]

Why is ATM-cell relay needed? At the local area level, shared-medium LANs operating at the 10 Mb/s range, such as Ethernet and token ring, are simply not capable of supporting a large number of simultaneous users all performing video or multimedia functions. Even assuming that a video window on a high-end workstation required only 1.5 Mb/s, a multipoint-to-multipoint conference may require four or eight video windows; in addition, users may also need to transmit large data files to various conferees as part of the interaction (for example, in cooperative work situations). In a 10 Mb/s LAN with 20 users (as an example), each user gets on average only one-twentieth of the bandwidth, that is, 0.5 Mb/s. However, the collision mechanism and the other higher layer protocols may use up 30 percent of the useful bandwidth, leaving each user with only 0.3 Mb/s. Worse yet, the random access techniques introduce variable delays, negatively affecting "isochronous" (also known as stream) services such as video and voice, which cannot be buffered beyond a few milliseconds.

With ATM techniques each workstation obtains a physical link operating (for now) at 155 Mb/s. The workstation can then partition this "pipe" into a number of virtual channels each with dedicated bandwidth to support voice, video, and data (see Fig. 9.2). Above the data link layer, traditional protocols such as TCP/IP can be used if desired. However, some "adaptation" function will be required, as alluded to earlier.

ATM-based wiring hubs allow each workstation, connected in a star arrangement, to have a dedicated (not shared) channel operating at 155 Mb/s. Short unshielded twisted pairs will be used, although a fiber medium can also be used if desired, particularly for

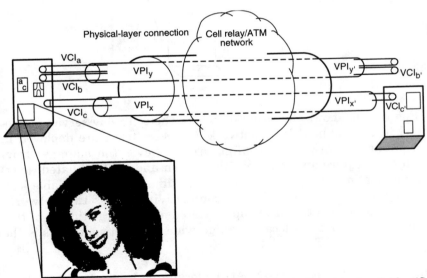

Figure 9.2 Multiple virtual circuits supported by ATM, including virtual path identifier (VPI) and virtual channel identifier (VCI). VCI_a and VCI_b represent two of the possible values of VCI within the VP link with the value VPI_y; VPI_x and VPI_y refer to two of the possible values of VPI within the physical-layer connection; VCI_c is another virtual channel. *Notes:* (1) VCIs and VPIs are only locally significant—hence a new VPI/VCI label may be used at the receiving end (shown here as VCI_a, etc.). (2) A multimedia application supported by a workstation could use a single VCI or multiple VCIs.

long reach. The physical layer protocols are based on SONET, and the data link layer is based on ATM protocols and related adaptation protocols. In addition to these data path protocols, protocols are also required for signaling, that is, to communicate supervisory requests to the wiring hub.

Is it real? It is likely that, by the end of 1993, there will be half a dozen ATM hub vendors and half a dozen ATM workstation vendors. Over 275 companies have recently joined the ATM Forum, which is an organization whose goal is to expedite and facilitate the introduction of ATM-based services. About two dozen vendors had announced firm equipment plans by publication time. Figure 9.3a depicts the rapid transition in workstation connectivity in the next year or so. The PC-workstation cards shown are expected to become available for about $1000 per port, although the initial cost was in the $1500-to-$3000 range. Figures 9.3b and 9.3c are illustrative of the type of equipment vendors are pursuing.

Some equipment vendors are building stand-alone switches; other are adding switching capabilities to their hubs and at the same time are developing ATM adapter cards for workstations to allow them to connect to the hub. Some are also working on bridge-router cards for ATM hubs that enable Ethernet LANs to connect to the ATM.[16]

It appears that in the United States the route to fast-packet services will be via ATM switch platforms; these support frame relay or cell relay interfaces on the access side and ATM-cell relay on the backbone side. In fact, several carriers are planning to use ATM as

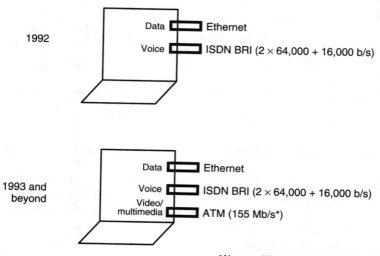

*Also possible: 1.5 Mb/s, 45 Mb/s

Figure 9.3a Emergence of the ATM-configured workstation.

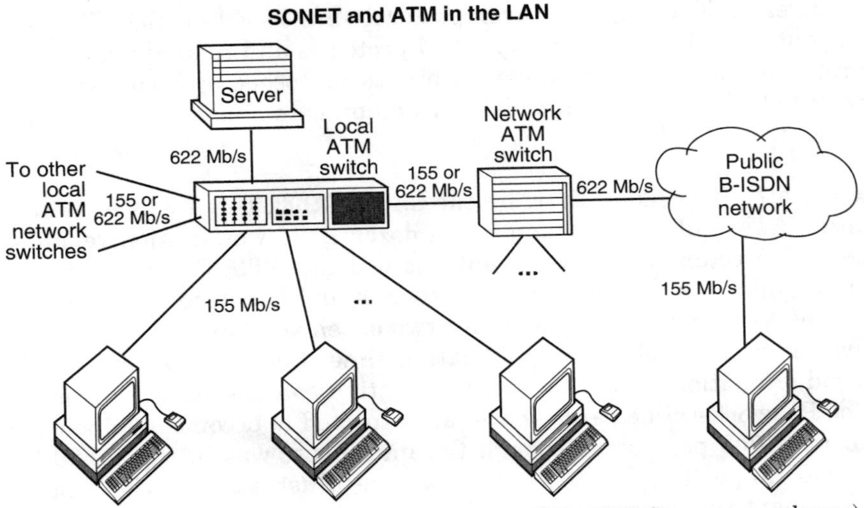
SONET and ATM in the LAN
Server
622 Mb/s
Local ATM switch
Network ATM switch
To other local ATM network switches
155 or 622 Mb/s
155 or 622 Mb/s
622 Mb/s
Public B-ISDN network
155 Mb/s
155 Mb/s
...
...

Figure 9.3*b* Illustrative premises of ATM equipment (from PMC-Sierra press releases).

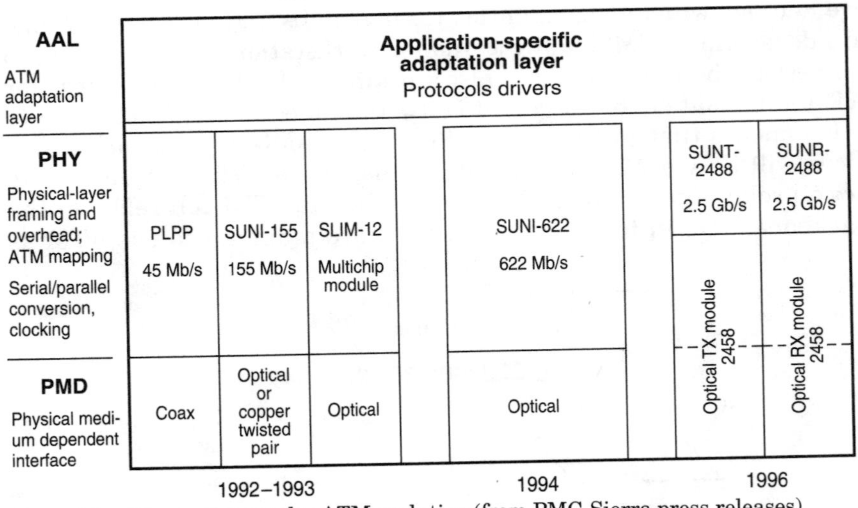
AAL
ATM adaptation layer
PHY
Physical-layer framing and overhead; ATM mapping
Serial/parallel conversion, clocking
PMD
Physical medium dependent interface
Application-specific adaptation layer
Protocols drivers
PLPP
45 Mb/s
SUNI-155
155 Mb/s
SLIM-12
Multichip module
SUNI-622
622 Mb/s
SUNT-2488
2.5 Gb/s
SUNR-2488
2.5 Gb/s
Optical TX module 2458
Optical RX module 2458
Coax
Optical or copper twisted pair
Optical
Optical
1992–1993
1994
1996

Figure 9.3*c* Illustrative vendor ATM evolution (from PMC-Sierra press releases).

a common platform to support frame relay, SMDS, and cell relay service.

Early ATM equipment or service entrants are shown in Table 9.1.[15–18] Industry proponents "expect to see Fortune 1000 users passing the majority of their LAN-to-WAN traffic through premises-based ATM switches by 1997."[19] Approximately 50 percent of the ATM traffic in these companies is expected to be in support of LAN interconnection, for LANs serving traditional business applications; the other 50 percent of the traffic is expected to be split fairly evenly among

TABLE 9.1   Partial List of Early Entrants in the ATM Equipment Market

WAN telecommunication and PBX vendors	DSC Communications Corp.; Fujitsu Network Switching of America; InteCom Inc.; NEC America; Siemens Stromberg-Carlson
Hubs and routers	Adaptive Corp.; ADC Fibermux; Fore Systems; Gandalf Systems Corp.; Hughes LAN; Newbridge Networks Inc.; SynOptics; Ungermann-Bass Inc.; Wellfleet Communications
Premises switches	Adaptive Corp.; Fore Systems; Frame Relay Technologies Inc.; Premisys Communications Inc.; PMC-Sierra; TRW Electronic Systems
Workstation cards	Adaptive Corp.; SynOptics
Other*	Ascom Timeplex; AT&T; Clearpoint Research Corp.; David Systems Inc.; Ericsson Business Communications; SynOptics

*Product plans unclear at publication time.

applications supporting real-time video, imaging, real-time voice, and multimedia.

9.2.2   What is ATM?

ATM entails the transport and switching of cells. A *cell* is a block of information of short fixed length which is comprised of an overhead section and a payload section. *Asynchronous time-division multiplexing* (ATM) is a multiplexing technique in which a transmission capability is organized in undedicated slots filled with cells supporting a user's instantaneous need. *ATM* is a transfer mode in which the information is organized into cells; it is asynchronous in the sense that the recurrence of cells containing information from an individual user is not necessarily periodic.[8,20,21]

Just as in traditional packet switching or frame relay, data is sent between two points—not over a dedicated, physically owned facility but over a shared facility comprised of virtual channels. Each user is assured that although other users or other channels belonging to the same user may be present, the user's data can be reliably, rapidly, and securely transmitted over the network in a manner consistent with the subscribed quality of service. The user's data is associated with a specified virtual channel. It must be noted that this type of sharing is not the same as a random-access technique where there are no guarantees of how long it can take for a data block to be transmitted.

In ATM, a *virtual channel* (VC) is used to describe unidirectional transport of ATM cells associated by a common unique identifier value, called *virtual channel identifier* (VCI). Even though a channel is unidirectional, the channel identifiers are assigned bidirectionally. The bandwidth in the return direction may be assigned symmetrically or asymmetrically, or it could be zero. *Virtual path* (VP) is used to describe unidirectional transport of ATM cells belonging to virtual channels that are associated by a common identifier value, called *virtual path identifier* (VPI). Refer again to Fig. 9.2. VPIs are viewed by some as a mechanism for hierarchical addressing. The VPI/VCI address space allows in theory up to 16 million virtual connections over a single interface; however, most vendors are building equipment supporting a minimum of 4096 channels on the user's interface. Note that these labels are only locally significant (at a given interface). They may undergo remapping in the network; however, there is an end-to-end identification of the user's stream so that data can flow reliably (e.g., path: VCI2 → VCI22 → VCI222; the sender labels cells with VCI = 2 and the receiver looks for cells with label VCI=222). Also note that on the network trunk side more than 4096 channels per interface are supported.

With cell relay, information to be transferred is converted into fixed-size cells. A cell consists of an information field (48 octets) and header (5 octets). The function of the header is to identify cells belonging to the same VC. Cells are identified and switched by means of the label in the header. Cell relay service allows for a dynamic transfer rate, specified on a per-call basis. Transfer capacity is assigned by negotiation and is based on the source requirements and the available network capacity. Cell sequence integrity on a virtual channel connection is preserved by ATM.

Now that we have described how cells can be identified, let us take a closer look at the protocol stack involved in ATM communication. Figure 9.4 illustrates this stack and shows the functions at each layer. This logical model is composed of a user plane, a control plane, and a management plane (not shown in Fig. 9.4). The *user plane,* with its layered structure, provides for user information transfer. Above the physical layer, the ATM layer provides information transfer for all services; the AAL provides service-dependent functions to the layer above the AAL. The control plane also has a layered architecture and performs the call control and connection functions. The layer above the AAL in the control plane provides call control and connection control. It deals with the signaling necessary to set up, supervise, and release calls and connections. The management plane provides network supervision functions. It provides two types of functions: layer management and plane management. Plane management performs management functions related to a system as a whole and provides coordination among all planes. Layer management performs management functions relating to resources and parameters residing in its protocol entities.

Physical layer functions.   The physical layer consists of two *logical* sublayers: the physical medium (PM) dependent sublayer and the

Control plane		User plane
Q.93B		Higher layers (e.g., TCP/IP and related protocols, or OSI)
ATM adaptation (AAL 5)	ATM adaptation	• Convergence (CS) • Segmentation and reassembly (SAR)
Asynchronous transfer mode		• Generic flow control • Cell header generation/extraction • Cell multiplex/demultiplex
Transmission convergence		• Cell rate decoupling • HEC header sequence generation/verification • Cell delineation • Transmission frame adaptation • Transmission frame generation/recovery
Physical medium dependent		• Bit timing • Physical medium

Figure 9.4 ATM communication protocol stack.

transmission convergence (TC) sublayer. PM includes only physical medium-dependent functions. It provides bit transmission capability, including bit transfer, bit alignment, line coding, and electrical-optical conversion. TC performs functions required to transform a flow of cells into a flow of information (i.e., bits) which can be transmitted and received over a physical medium. TC functions include (1) transmission frame generation and recovery, (2) transmission frame adaptation, (3) cell delineation, (4) header error control (HEC) sequence generation and cell header verification, and (5) cell rate decoupling.

Cell rate decoupling includes insertion and suppression of idle cells, in order to adapt the rate of valid ATM cells to the payload capacity of the transmission system. The transmission frame adaptation function performs the actions which are necessary to structure the cell flow according to the payload structure of the transmission frame (transmit direction) and to extract this cell flow out of the transmission frame (receive direction). The transmission frame requires (in the United States) SONET or DS3 envelopes. Cell delineation prepares the cell flow in order to enable the receiving side to recover cell boundaries. In the transmit direction, the ATM cell stream is scrambled. In the receive direction, cell boundaries are identified and confirmed and the cell flow is descrambled.

The HEC mechanism covers the entire cell header, which is available to this layer by the time the cell is passed down to it. The code used for this function is capable of either single-bit correction or multiple-bit error detection. The transmitting side computes the HEC field value.

The service data units crossing the boundary between the ATM layer and the physical layer constitute a flow of valid cells. The ATM layer is unique, that is, independent of the underlying physical layer. The data flow inserted in the transmission system payload is physical medium-independent and is self-supported; the physical layer merges the ATM cell flow with the appropriate information for cell delineation, according to the cell delineation mechanism.

The transfer capacity at the user-network interface (UNI) with an ATM network is 155.52 Mb/s, with a payload capacity of 149.76 Mb/s. Since the ATM cell has 5 octets of overhead, the 48-octet information field equates to a maximum of 135.631 Mb/s of actual user information. A second UNI interface is defined at 622.08 Mb/s, with the service bit rate of approximately 600 Mb/s. Access at these rates requires a fiber-based plant. Other UNIs are also being contemplated in the United States at the DS1 and DS3 rate. The DS1 UNI is discussed in the context of an electrical interface (T1); so is the DS3 UNI.

ATM layer functions.   Connection identifiers are assigned to each link of a connection when required and are released when no longer needed. ATM offers a flexible transfer capability common to all services, including connectionless services. The transport functions of the ATM layer are subdivided into two levels: the virtual channel level and the virtual path level. The transport functions of the ATM layer are independent of the physical layer implementation. The label in each ATM cell is used to explicitly identify the VC to which the cells belong. The label consists of two parts: the VCI and the VPI. A VCI identifies a particular VC link for a given virtual path connection. A specific value of VCI is assigned each time a VC is switched in the network. With this in mind, a VC can be defined as a unidirectional capability for the transport of ATM cells between two consecutive ATM entities where the VCI value is translated. A VC link is originated or terminated by the assignment or removal of the VCI value.

The functions of ATM include (refer again to Fig. 9.4):

Cell multiplexing and demultiplexing.   In the transmit direction, the cell multiplexing function combines cells from individual VPs and VCs into a noncontinuous composite cell flow. In the receive direction, the cell demultiplexing function directs individual cells from a noncontinuous composite cell flow to the appropriate VP or VC.

Virtual path identifier and virtual channel identifier translation.
This function occurs at ATM switching points and cross-connect
nodes. The value of the VPI and VCI fields of each incoming ATM
cell is mapped into a new VPI and VCI value. This mapping function could be null.

Cell header generation and extraction. These functions apply at
points where the ATM layer is terminated. In the transmit direction, the cell header generation function receives cell payload information from a higher layer and generates an appropriate ATM cell
header except for the HEC sequence. In the receive direction, the
cell header extraction function removes the ATM cell header and
passes the cell information field to a higher layer.

ATM adaptation layer. Additional functionality on top of the ATM
layer (i.e., in the ATM adaptation layer) must be provided to accommodate various services. The ATM adaptation layer enhances the services provided by the ATM layer to support the functions required by
the next higher layer. The boundary between the ATM layer and the
AAL corresponds to the boundary between functions supported by the
contents of the cell header and functions supported by AAL-specific
information (also see Fig. 9.8). In general, the AAL-specific information is nested in the information field of the ATM cell.

Connections in an ATM network support both circuit mode and
packet mode (connection-oriented and connectionless) services of a
single medium or mixed media and multimedia. It services two types
of traffic: constant bit rate (CBR) and variable bit rate (VBR). CBR
transfer rate parameters for on-demand services are negotiated at
call setup time. (Changes to traffic rates during the call may eventually be negotiated through the signaling mechanism; however, initial
deployments will not support renegotiation of bit rates.) CBR transfer
rate parameters for permanent services are agreed upon with the carrier from which the user obtains service. VBR services are described
by a number of traffic-related parameters (minimum capacity, maximum capacity, etc.). The AAL protocols are used to support these different connection types.

The AAL performs functions required by the user, control, and
management planes and supports the mapping between the ATM
layer and the next higher layer. Note that a different instance of the
AAL functionality is required in each plane. The AAL supports multiple protocols to fit the needs of the different AAL service users; hence,
the AAL is service-dependent. Namely, the functions performed in the
AAL depend upon the higher-layer requirements. The AAL isolates
the higher layers from the specific characteristics of the ATM layer by
mapping the higher-layer PDUs into the information field of the ATM

cell and vice versa. The AAL entities exchange information with the peer AAL entities to support the AAL functions.

The AAL functions are organized in two logical sublayers, the convergence sublayer (CS) and the segmentation and reassembly sublayer (SAR). The function of CS is to provide the AAL service at the AAL service access point. This sublayer is service-dependent. The functions of SAR are (1) segmentation of higher-layer information into a size suitable for the information field of an ATM cell and (2) reassembly of the contents of ATM cell information fields into higher-layer information.

In order to minimize the number of AAL protocols, a service classification is defined based on the following three parameters: (1) timing relation between source and destination (required or not required), (2) bit rate (constant or variable), and (3) connection mode (connection-oriented or connectionless). Other parameters such as assurance of the communication are treated as quality of service parameters and therefore do not lead to different service classes for the AAL. Five classes of application are then defined as follows:

Class A: Timing required, bit rate constant, connection-oriented

Class B: Timing required, bit rate variable, connection-oriented

Class C: Timing not required, bit rate variable, connection-oriented

Class D: Timing not required, bit rate variable, connectionless

Class X: Unrestricted (bit rate variable, connection-oriented or connectionless)

Three AAL protocols have been defined in support of different applications; these are AAL type 1, AAL type 3/4, and AAL type 5. It appears that the computer communication community (building LAN and multiplexing equipment) will use AAL type 5. Additionally, the ATM service likely to be available first (and the one supported by evolving computer equipment) is the one related to Class X, which is known as broadband connection-oriented bearer service—Class X (BCOB-X). A simpler name is cell relay service.

9.2.3   ATM cell layout

The UNI ATM cell format specified in ITU-T recommendation I.361 consists of a 5-octet header and a 48-octet information field (also called payload). The structure of the ATM cell across the user-network interface is shown in Fig. 9.5 (the network-node interface cell is identical to the UNI cell except that the VPI occupies the entire first octet, rather than just bits 5 through 8). For the UNI, 24 bits are available for rout-

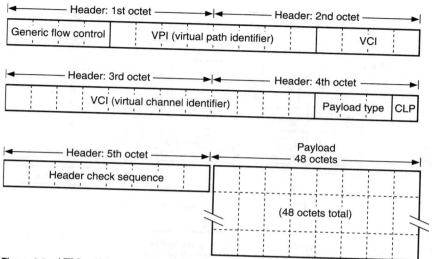

Figure 9.5 ATM cell layout. CLP = cell loss priority. *Notes:* (1) The actual number of routing bits in the VPI and VCI fields is negotiated between the user and the network; this number is determined on the basis of the lower requirement of the user or the network. (2) This cell represents the cell at the user-network interface. (3) Overhead of AAL and related protocols must fit inside the ATM payload.

ing: 8 bits for the VPI and 16 bits for the VCI; 3 bits are available for payload type identification. This is used to provide an indication of whether the cell payload contains user information or network information. In user information cells the payload consists of user information and service adaptation function information. In network information cells the payload does not form part of the user's information transfer. If the cell loss priority (CLP) is set (CLP value is 1), the cell is subject to discard, depending on the network conditions. If the CLP is not set (CLP value is 0), the cell has higher priority. The header error control field consists of 8 bits and is used for error management of the header.

Figure 9.6 shows the protocol apparatus required in the user's equipment (e.g., workstation). Note that two stacks must be implemented in order to obtain VCs on demand (this is also known as SVC service). With this capability, the user can set up and take down multiple connections at will. Initially it is likely that a PVC service will be available. In this mode, the control plane stack is not required and the desired connections are established at service initiation time and remain active for the duration of the service contract. Also note that AAL functions (SAR and CS) must be provided by the user equipment. (In a class A service, not shown in the figure, the AAL functions are provided by the network.) Also, the user equipment must be able to assemble and disassemble cells (i.e., run the ATM protocol).

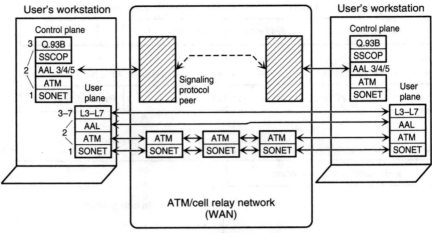

Figure 9.6 End-user workstation protocol requirements to access an SVC ATM network. SSCOP = service-specific connection-oriented protocol.

9.3   Local ATM Technology

Local ATM refers to the application of ATM technology to premises-based communication.

9.3.1   Local ATM environment

The environment to which LATM is directed is composed of four entities:

- User devices (including servers and hosts)
- User-owned ATM switches (possibly in hubs)
- Internetworking devices such as routers
- The public ATM/B-ISDN/cell relay network

Figure 9.7, from the network-compatible LATM specification,[1] depicts this environment from a topological perspective. Devices connect to ATM switches over a point-to-point link. This interface is referred to as I_1. If desired, devices can have physical connectivity to multiple-premises ATM switches; however, each interface has a unique address. The physical connection between premises switches is described by interface I_2. The interface between the local ATM switch and the public B-ISDN network is called I_3 (a device connecting directly the B-ISDN network, without the services of a local ATM switch, also employs interface I_3). This implies that a device could

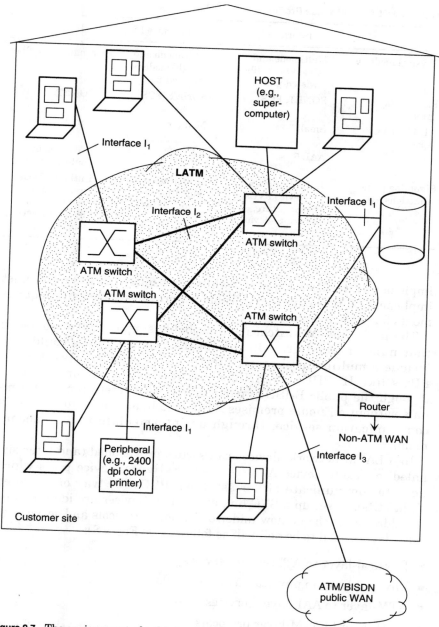

Figure 9.7 The environment of a third-generation LAN.

TABLE 9.2   LATM Interface Profiles

	Interface I_1	Interface I_2	Interface I_3
Physical medium	Multimode and single-mode fiber, twisted-pair	Multimode and single-mode fiber, twisted-pair	Single-mode fiber
Physical layer protocol	SONET, FCS	SONET, FCS	SONET
ATM VPI/VCI space	Small	Medium	Large
AAL	AAL 5	AAL 5	AAL used by public network
Signaling plane functions	Initially, device-to-device PVCs; in the future, device-to-device SVCs	Switch-to-switch signaling capabilities	Initially, PVCs; in the future, signaling used by public network

implement interfaces I_1 and I_3, while a premises ATM switch must implement all three interfaces. Table 9.2 provides some early requirements for these interfaces.

The initial LATM switches will probably support 100 or so users; that number could grow in the future. Some vendors could also include a multiprotocol router function (i.e., process network layer PDUs like IP PDUs). WAN communication is supported either through the public B-ISDN, particularly for discrete remote terminals without a remote-premises ATM switch, or over a non–B-ISDN communication service, through an intermediate router mechanism.

In a LATM network, devices can secure provisioned (and later signaled) device-to-device ATM connection. Hence, device pairs which need to communicate either establish PVCs (by way of assigned VPIs/VCIs), or set up a SVC link as needed. Devices are identified by an addressing scheme now under definition. Protocols and services at three layers need to be defined, as follows (see Fig. 9.8):

- Physical-layer to ATM-layer services
- Physical-layer to physical-layer protocols
- ATM-layer to AAL-layer services
- ATM-layer to ATM-layer protocols
- AAL-layer to higher-layer services
- AAL-layer to AAL-layer protocols

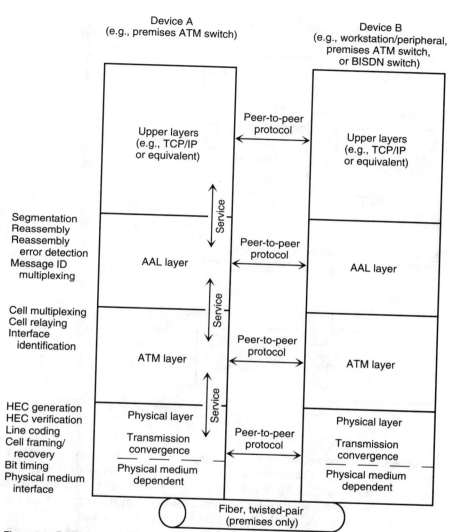

Figure 9.8 LATM protocols.

Figure 9.9 can be utilized in the sequel to position the various service access points under discussion.

9.3.2 Physical-layer to ATM-layer services

The function of the physical layer is to move incoming PHY SDUs (i.e., cells) between two peer systems supporting the same physical layer. As is the case with any layer, the PHY SDU is prepended with

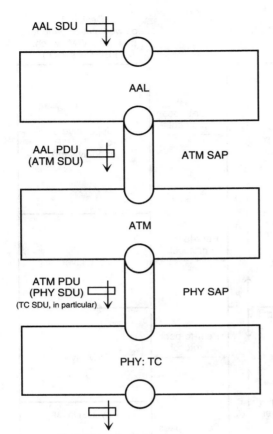

Figure 9.9 Pertinent SAPs.

layer-specific protocol control information (PCI) in order to transmit it to the peer entity. Transmission is bidirectional and symmetric, initially supporting 155.52 Mb/s (622 Mb/s later). The PCI includes transmission overhead like HEC, performance monitoring, and alarm bits. As seen in Table 9.2, both multimode (50/125 μm and 62/125 μm) and single-mode fiber is supported in the premises wiring; however, only single-mode is supported over the public network UNI. Twisted-pair wiring is also supported locally. Only cells which have headers without bit errors are delivered to the higher layer with an *indication primitive* over the PHY-SAP.

The physical layer comprises a *transmission convergence* (TC) sublayer residing over a *physical medium-dependent* (PMD) sublayer.

PMD functions include encoding of the bits into an appropriate electrical or optical signal. Bit transmission, symbol alignment, and timing extraction functions are supported in this sublayer. Timing for the SONET TC is (initially) 155.52 Mb/s ± 20 ppm at I_3 and 155.52

Mb/s $\pm$ 50 ppm at I_1 and I_2. The PMD also specifies the type of fibers, the transmitter-receiver wavelength, the receiver sensitivity, and the power budget (attenuation and dispersion) among parameters. The optical connector type and its performance characteristics are also specified.

TC provides convergence function to the SONET structure (adapts the cell for transmission over the transmission structure). The TC sublayer generates and recovers transmission frames. Transmission frames are generated by adding the PCI to the cells (PHY SDUs) received across the SAP. These cells are then accommodated within the payload structure of SONET or FCS. Data transmission functions of TC facilitate the movement of bits over the physical channel. Data reception functions perform the inverse process of data transmission. Cell delineation functions identify the cell boundaries within a bit stream. HEC generation functions calculate the HEC over the PHY SDU's header. HEC processing functions allows the receiving peer entity to determine cell header errors and possible correct single-bit errors; the HEC field can also be used for cell framing recovery. The TC sublayer also provides line scrambling and descrambling to enhance clock recovery, minimize false cell framing, and enable the user to submit long transparent sequences of 0s (generator polynomial: $x^{43} + 1$). There are also performance monitoring functions.

Two TC approaches are proposed for support by LATM: SONET and the block-coded scheme used in the FCS (see Chap. 10). Under the SONET TC, ATM cells are mapped into the SONET synchronous payload envelope (SPE) (but, clearly not in the overhead octets). Since cells "slide" within the SONET frame (the latter being 2340, not an integral multiple of the 53-octet cell), a mechanism is required to establish the beginning of a cell. This is done by indicating in the overhead's H4 field the offset to the closest cell which follows (this counter is ≤ 52).

The "user plane-signaling plane" operation of the TC sublayer of the PHY layer is as follows. On a UNITDATA.request from the ATM layer, it generates the HEC and then the necessary physical transmission frame. The PMD is responsible for actual transmission. At the receiving end, the TC sublayer extracts the SDU from the received PDU, and if the HEC mechanism shows a valid header, a UNITDATA.indication to its ATM layer is provided. (Refer back to the OSIBRM description in Chap. 2.)

The PHY layer also has a SAP to the "management plane." Indications for events as loss of signal, loss of frame, loss of cell delineation, uncorrectable HEC, far-end failure, and alarm condition are provided.

9.3.3 Physical-layer to physical-layer protocols

We identify this protocol with the SONET structure in effect between peer TC sublayers. SONET at STS-3c provides cell transport at 149.760 Mb/s and an actual payload throughput of 135.632 Mb/s.

In the STS-3c frame (270 octets × 9 octets = 2430 octets per 125 μs) only the following transmission overhead octets are used (these reside in the first 9 octets × 9 octets): A1, A2, B1, C1 (in the section overhead), B2, H1, H2, H3, K2 and Z2 (in the line overhead). In the path overhead, only octets J1, B3, C2, G1, and H4 are used.

In LATM applications, the section overhead supports the following functions: frame alignment (A1, A2), frame identifier (C1), and sec-

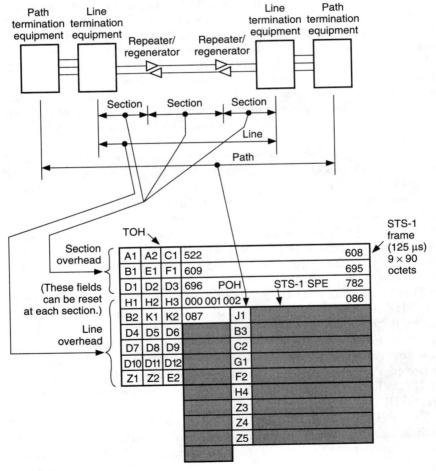

Figure 9.10 SONET frame.

tion error monitoring (B1). Line overhead supports the following functions: line error monitoring (B2), pointer and concatenation (H1 and H2), and line performance monitoring and alarm indication (K2, Z2). The path overhead supports trace (J1), path error monitoring (B3, G1), and cell offset (H4). Figure 9.10 depicts the SONET frame for STS-1. Figure 9.11 shows the carriage of ATM frames within the underlying SONET frame. See Minoli[8] for an extensive treatment of SONET.

9.3.4 ATM-layer to AAL-layer services

The ATM layers provide for the transfer of cells (ATM SDUs) in support of the AAL. Both a point-to-point and a point-to-multipoint transmission service are supported. Initially, the ATM connection needs to be an already-established PVC (with specified cell-loss ratio, throughput, cell delay, and other transmission performance parameters); in

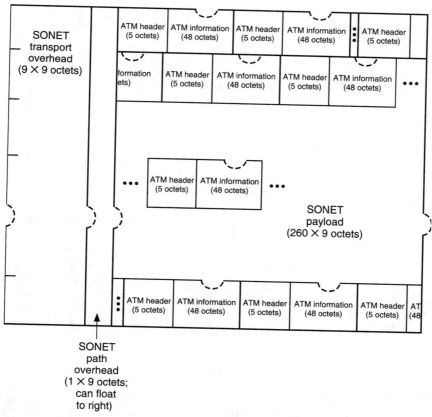

Figure 9.11 Mapping of ATM cells into SONET frame.

Phase II, using the 1993 ATM Forum UNI specification, the ATM connections will be signaled as required. Local ATM switches do not need to monitor or enforce throughput (measured as the number of cells per time unit), although public B-ISDN switches will. Proper sequencing of cells is provided except possibly for gaps generated by lost cells (no retransmission mechanism is supported) or corrupted cells (which are not relayed by an intermediate node which detects the corruption by means of the HEC). Figure 9.12 depicts the required implementation of the ATM layer, given different user arrangements.

Key functions of the ATM layer are shown in Table 9.3. Note that cells are switched or relayed from one VP to another VP (VP switch), or from one VC to another VC in the same or a different VP (VP/VC switch).

Over the ATM SAP, two primitives are supported: ATM-DATA.request and ATM-DATA.indication. The ATM-DATA.request primitive initiates the transfer of an ATM SDU to the peer entity in

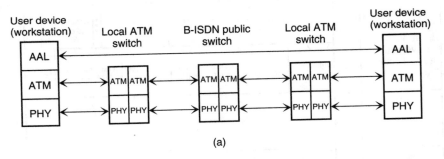

(a)

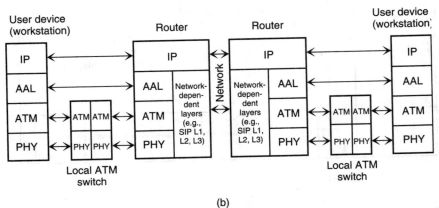

(b)

Figure 9.12 Termination of ATM and other layers: (*a*) Using B-ISDN; (*b*) using a router over a non–B-ISDN network. (*Note:* Here the local ATM switch is shown as reconstituting a cell. This would be the case, for example, if the VCI/VPI were changed by the switch.)

TABLE 9.3   Key Functions of the ATM Layer

Cell transmission	Issues a data transfer request to the physical layer to initiate transmission of data to a remote peer.
Cell reception	Accepts a data reception indication from the physical layer to initiate data acceptance from a remote peer.
Cell relaying	Forwards an incoming cell (with a given VPI_x/VCI_y) to another entity (with a given VPI_u/VCI_v). This function is done only at intermediate nodes. Nodes must be able to translate the supported receive VPI/VCI combinations into the supported transmit VPI/VCI combinations.
Cell multiplexing	Aggregates cells from several VC/VPs into a composite flow of cells. This is done where there are multiple devices over the same UNI interface. Individual connections are identified by unique VPI/VCI labels.
Cell demultiplexing	Differentiates incoming cells over a UNI and delivers them to the appropriate end point. Utilizes the VPI, VCI, and PTI fields.
Cell rate decoupling	This function inserts "unassigned" cells into the transmission flow to create a continuous stream of cells. Opposite function is undertaken at receiving end.
Delay priority processing	Schedules transmission according to priority requirement.
Cell loss priority marking	Enables ATM entity to mark each cell as to its required cell loss priority.
Cell loss priority reduction	Allows intermediate system to modify the CLP of any cell contributing to an input exceeding the allowed peak rate.
Cell rate pacing	Supports a mechanism limiting the cell input rate (cells per unit time).
Cell rate transmission when exceeding pacing rate	This function allows the ATM entity to transmit at a rate higher than originally provisioned (via a signaling or reprovisioning mechanism).
Peak-rate enforcement	Allows monitoring of each connection to identify cells that are not in compliance with the agreed enforcement traffic descriptor. This may entail performing cell loss priority reduction or discarding noncomplying cells.
Explicit forward-congestion marking	Allows congested intermediate nodes to set an explicit forward notification indicator to inform destinations (impacts PTI field).
Explicit forward-congestion indication	Allows ATM entity to inform higher layers of a state of congestion.
Cell payload type marking	Allows marking of cells (e.g., management cell: PTI = 110; user cell, no congestion encountered, last cell: PTI = 001; user cell, no congestion encountered, not-last-cell: PTI = 000; user cell, congestion encountered, last cell: PTI = 011; etc.).
Cell payload type differentiation	Allows differentiating cells based on the PTI.
Generic flow control	Supports standardized local functions (e.g., flow control when there are multiple devices sharing a single UNI). It supports two modes: "uncontrolled access" means there are no restrictions on the host; "controlled access" means that hosts are expected to modify their transmissions according to the content of the field.

the target device. In addition to the data, a cell priority parameter is passed down to the ATM layer. The ATM-DATA.indication primitive alerts the AAL layer to the arrival of an ATM SDU. In addition to the data, a congestion indication is set if the cell passes through a portion of the network experiencing congestion.

9.3.5   ATM-layer to ATM-layer protocols

One can view the protocol as the mechanism embodied in the cell frame structure, plus additional procedures to deal with exceptional cases (e.g., header missing a field, encoding of a specific field not allowed, etc.).

9.3.6   AAL-layer to higher-layer services

The function of the AAL is to provide data link layer capabilities to the higher layers, utilizing the services of the ATM layer. It accepts upper-layer (e.g., TCP/IP) PDUs (varying from 0 to 65,535 octets) and transmits them to a peer AAL with error detection. The protocol aims at being efficient and simple. This protocol is also known as *data transfer AAL* or *AAL type 5.*[14] One of the key functions of this layer is segmentation and reassembly of data into cells.

This layer accepts an AAL-UNITDATA.request from the upper layer and issues an AAL-UNITDATA.indication to the upper layer. A variety of parameters needs to be passed with these primitives as part of the service request. There is a SAP to layer management, which functions to create and remove AAL connections, to report errors, to set connection parameters, and for other functions. Again, a variety of parameters need to be passed with these primitives as part of the service request. Figure 9.13 depicts the format of the AAL-PDU. The two-octet control field is reserved for future use. The length field indicates the length in octets of the user data field.

The AAL expects the ATM layer to provide the multiplexing and transport of segments of 48 octets between AAL entities. This is done through an ATM-DATA.request and ATM-DATA.indication. The ATM-DATA.request primitive includes one data segment, a loss-priority indication, and an SDU-type. The ATM-DATA.indication includes a data segment, the SDU-type (0 = continuation, 1 = last), and any indication of congestion.

Six key AAL functions are:

1. *AAL-PDU generation by transmitter AAL.* This function adds the appropriate padding to make the entire AAL PDU (including the length field, control field, and CRC-32) divisible by 48. It also concatenates these fields to the PDU.

User data (up to 65,535 octets)	Pad (to align user data field to 48-octet boundary)	Length (2 octets)	Control (2 octets)	CRC-32

Figure 9.13 LATM AAL-PDU format.

2. *CRC generation by transmitter AAL.*

3. *Segmentation by transmitter AAL.* This function generates successive blocks of 48 octets (beginning with the most significant octet). All segments except the last one are submitted to the ATM layer (with the AAL-DATA.request) with the ATM SDU-type set to zero.

4. *Reassembly by receiver AAL.* This function enables reconstruction of the AAL PDU from the segments received via ATM.indication primitives. Successive blocks are appended until one (supposedly the last one) with an SDU-type bit set to 1 for end of message (EOM). If there are no problems (lost cells, dropped cells, delayed cells, foreign or misrouted cells) the PDU can be reconstructed; otherwise error procedures must be initiated. The determination of whether a problem was encountered or not is achieved using the CRC mechanism.

5. *CRC validation.* This receiver function computes a CRC-32 over the entire "working" AAL PDU (by the concatenation of incoming ATM payloads). If the test passes, it follows that the "working" PDU has been correctly assembled.

6. *AAL-SRU recovery.* This receiver function identifies the AAL-SDU boundaries within the AAL PDU using the length field (to identify which octets, if any, comprise the pad).

9.3.7   AAL-layer to AAL-layer protocol

Procedures associated with the functions described in the previous sections (protocol state machine, error conditions, etc.) have been defined. The protocol supports "uninsured" data transfer, but does detect transmission errors. No addressing is supported since the VPI/VCI capability of the ATM layer does this.

As AAL type 5 was being defined in B-ISDN standards bodies, the protocol was partitioned into an upper sublayer (service-specific protocol—SSP), which can be null (in most instances), and a lower sublayer called "common part."

9.3.8   Signaling

As discussed elsewhere, communication can take place in a connectionless or connection-oriented mode. Each technique has niches where it is the best technology. For example, voice communication over the public switched network has used the connection-oriented approach of call setup, information transfer, and call tear-down for over a century. By contrast, process-control environments may be better off with connectionless communication. Consider, for example, several hundred devices providing perimeter security for a nuclear power plant or sensors on a battleship. Typically, these sensors only need to send a packet of information on a time-driven basis (say every 5 seconds), and additionally on an event-driven basis (say an intrusion or malfunction). In this case it would not make sense to provide dedicated bandwidth to many remote sensors (because of the low traffic volume—real or PVC-based channels) or a switched service (owing to the possibly detrimental delay in sending the information to a command center and receiving data for remedial action).

Connectionless techniques are ideal for maximizing bandwidth efficiency by providing a reliable multiplexing method. In recent years, bandwidth has become increasingly more available and more affordable. Some feel that bandwidth efficiency traded off for simplicity and user equipment is no longer the strategy which maximizes carrier revenues. By way of analogy, in the 1950s and 1960s people spent a lot of time optimizing software code so that it would utilize less RAM; today, with the increased cost of programming resources and the decreased cost of memory, this optimization is no longer economical.

Many user applications are intrinsically connection-oriented. Hence, while a system for process control, a system supporting automatic teller machines, point-of-sale credit authentication, lottery ticket sales, and a network to poll a population may be intrinsically connectionless (short, single-packet transactions), many business applications are connection-oriented (i.e., they require the low-latency, low-interframe-delay delivery of thousands of related messages). For example, word processing (say, delivery of a 20-page memo), multimedia messaging (with voice and video), video conferencing, and periodic database synchronization of large distributed databases are ultimately connection-oriented across the API. As an analogy, the U.S. postal system works fine as it stands (it is a connectionless-like system, where users drop individual letters into the system, without having to preallocate—i.e., reserve with a telephone call—the services to be received). However, no one would want to send 1000 pages of a report to a given remote destination by sending 1000 individually enveloped and stamped letters!

The B-ISDN bearer service expected in the immediate future is a connection-oriented service; this is observable by noting that it is based on the VCI/VPI mechanism. In turn, ATM connections can be prespecified PVCs or can be SVCs. Traditional LANs are connectionless at the lower layers. This technology has satisfactorily met the *data transfer needs* over a LAN (although, as noted, some applications may in fact be taxing such a system because they are intrinsically connection-oriented). However, with the need to transfer voice and video messages of hundreds of information frames per second per user (e.g., T1 equates to $8000 \times 24/1000 = 192$ 1000-octet frames per second), the use of a connection-oriented fabric appears more reasonable.

Third-generation "gigabit" LANs developed to support high-population multimedia are reverting to relying on connection-oriented techniques based on B-ISDN, as described in the earlier sections. In this environment, pairs of users can establish PVCs or SVCs over which high-data-rate communication with high frame-to-frame autocorrelation can take place efficiently. PVC-based methods, however, become restrictive from a management perspective as the number of users increases. Hence, SVC-based methods are being developed.

In SVC-based LATM communication, two multimedia terminals requiring communication go through a call-setup phase, after which communication can take place. (Note again that this is not the case in a first- or second-generation LAN.) The design objective is to keep this setup phase as short as possible, say 50, or 100 ms; a long setup phase (say, 5 to 10 seconds) would impede communication. Signaling is required to communicate the user requests (call control and bearer-bandwidth control information) to a bandwidth manager, typically the (L)ATM switch. In addition these techniques are necessary when communicating over a B-ISDN WAN, as shown in Fig. 9.1e and 9.1f. Additional functions include multiparty conference coordination, flow control, etc.

Work aimed at developing LATM-based signaling was under way at press time in the ATM Forum and TS SG XI; this work was expected to be completed in August 1993.

9.3.9   Multiprotocol carriage over ATM

In a general multiplexed ATM VC, say a WAN link carrying LAN traffic (TCP/IP for first- or second-generation LANs) as well as other traffic, there is a need to identify which protocol is being carried. The Internet engineering task force has been working on multiprotocol interconnects. There are three approaches to accommodate connectionless traffic, identified in their Internet draft, *Multiprotocol Interconnect over ATM AAL 5*:[22,23]

1. In the LLC/SNAP encapsulation method, multiple protocols (e.g., NetWare IPX, XNS, IP, AppleTalk, and IEEE 802.5 bridged PDUs) may be carried over a single virtual channel connection. Protocol identification of the AAL5-SDU content is by use of an IEEE 802.2 LLC header, usually followed by an IEEE 802.1a SNAP header (see Chap. 3). This approach allows interworking with IEEE 802 LANs.

2. In the NLPID/SNAP method, multiple protocols may be carried over a single virtual channel connection. The ISO/IEC TR 9577 network layer protocol ID (NLPID), sometimes in conjunction with a SNAP header, is used for protocol identification. This approach allows ATM-frame relay internetworking.

3. In the null encapsulation, only one protocol is carried on a virtual channel connection. Protocol identification is done by means of Q.93B signaling. This approach conserves overhead (for example, it permits an IP packet containing a TCP ACK packet to consume exactly one ATM cell).

Resolution as to whether the IETF will keep all three methods or converge on a subset is expected by 1993.

9.4 FDDI Follow-on LAN Approach

Observers note that, "as a backbone FDDI has been a big disappointment. Current backbone implementations of FDDI have uncovered numerous limitations of the technology, which are inherent to the token-passing nature of the FDDI MAC."[24] When FDDI is used as a backbone for LAN interconnection, internetworking hardware required to support the necessary conversion between other LANs and the FDDI MAC must be provided at every wiring closet and equipment room in the network. "In the area of performance, FDDI lacks sufficient response speed to support many data communications applications," not to mention multimedia applications.[24] Observers note that "there is a growing concern that the deficiencies in FDDI make it handicapped as a backbone technology—especially when considering its hefty price tag."[24]

In response to these clear FDDI limitations in the age of data-intensive applications, ANSI's X3T9.5 committee has begun work on a multiservice standard for high-speed backbones, known as FDDI follow-on LAN (FFOL) approach. The specifications that are required to support FFOL (see Fig. 9.14) include:

- FDDI follow-on LAN—physical medium dependent (FFOL-PMD)
- FDDI follow-on LAN—physical layer protocol (FFOL-PHY)
- FDDI follow-on LAN—service multiplexer (FFOL-SMUX)
- FDDI follow-on LAN—asynchronous MAC (FFOL-AMAC)
- FDDI follow-on LAN—isochronous MAC (FFOL-IMAC)
- FDDI follow-on LAN—station management (FFOL-SMT)

Indications are that the bandwidth targeted by FFOL are STS-3c, STS-12c, and STS-48c. The goals of the effort are naturally similar to

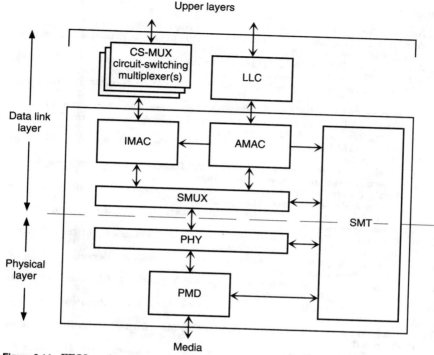

Figure 9.14 FFOL architecture.

those of LATM: scalable bandwidth, affordable plug-and-play, multi-vendor interoperability, seamless integration of LANs and WANs, support of LAN traffic, support of voice and video, and compatibility with EIA/TIA 568 building wiring standard.[24-26]

The SMUX being proposed is based on TDM rather than token methods, based on the recognition of the limitations imposed by microscopic bandwidth management at gigabit per second rates. With TDM, access to the medium (likely fiber) is granted during assigned intervals (time slots). Each node is assigned one or more slots, and delay-sensitive applications are assigned slots more often.[24-26] This could make delays several orders of magnitude lower than with FDDI. However, the MAC remains similar to a LAN MAC. This implies that achieving the FFOL "presents a significant, but not insurmountable challenge: defining a sufficiently robust, high-performance SMUX/MAC interface that accommodates the various incompatible payloads and word length." A sizable segment of the ATM Forum constituency believes that ATM is precisely such a mechanism, particularly with respect to interworking with WANs. Some see LATM as a more promising technology, also in the context of time-to-market considerations.

References

1. *Network Compatible ATM for Local Network Applications,* Phase 1, Version 1.0, April 1992, Anonymous FTP (Internet) at ftp.apple.com pub/latm/nclatm.ps or thumper.bellcore.com pub/latm/nclatm.ps.
2. J. B. Lyles and D. C. Swinehart, "The Emerging Gigabit Environment and the Role of Local ATM," *IEEE Communications Magazine,* April 1992, pp. 52 ff.
3. *ATM User-Network Interface Specification,* ATM Forum, Mountain View, Calif., 1992.
4. S. Kolodziej, "ATM Gains Supplier Interest," *Lightwave,* August 1992, pp. 1 ff.
5. D. Minoli, "ATM: The Future of Local and Wide Area Networks," *Network Computing,* October 15, 1992, pp. 128 ff.
6. D. Minoli, "ATM Protocols: Let's Get Technical," *Network Computing,* November 15, 1992, pp. 156 ff.
7. D. Minoli, "ATM and Cell Relay Concepts," Datapro Communications Series: Broadband Networking, Report #2880, 4/1992.
8. D. Minoli, *Enterprise Networking, Fractional T1 to SONET, Frame Relay to B-ISDN,* Artech House, Norwood, Mass., 1993.
9. C. E. Catlett, "In Search of Gigabit Applications," *IEEE Communications Magazine,* April 1992, pp. 42 ff.
10. N. K. Cheung, "The Infrastructure for Gigabit Computer Networks," *IEEE Communications Magazine,* April 1992, pp. 60 ff.
11. H. T. Kung, "Gigabit Local Area Networks: A Systems Perspective," *IEEE Communications Magazine,* April 1992, pp. 79 ff.
12. P. Newman, "ATM Technology for Corporate Network," *IEEE Communications Magazine,* April 1992, pp. 90 ff.
13. Special Issue on Multimedia Communications, *IEEE Communication Magazine,* May 1992.
14. Z. Wang and J. Crowcroft, "SEAL Detects Cell Misordering," *IEEE Network,* July 1992, pp. 8 ff.
15. E. Mier, "The Cell Switching Revolution," *Communications Week,* December 14, 1992, p. 61.
16. S. Girishankar, "Hughes Details ATM Product Plan," *Communications Week,* December 14, 1992, p. 29.
17. S. Girishankar, "Cabletron Working on ATM Switch, Interface," *Communications Week,* June 8, 1992, pp. 28 ff.
18. S. Girishankar, "New Fibermux Hub to Bolster Backbone Speeds with ATM," *Communications Week,* June 8, 1992, pp. 5 ff.
19. E. Mier, "The Vendor Perspective on ATM," *Communications Week,* December 14, 1992, p. 61.
20. D. Minoli, "Broadband Integrated Services Digital Network," Datapro Communications Series: Broadband Networking, Report #2890, 4/1992.
21. D. Minoli, "Third-Generation LANs," UNIX Expo 92 Proceedings, Bruno Blemheim Inc., Fort Lee, N.J., 1992.
22. Internet Engineering Task Force, *Multiprotocol Interconnect over ATM AAL 5,* Anonymous FTP, nic.ddn.mil, nnsc.nsf.net, ftp.nisc.sri.com, July 1992.
23. D. Grossman, T1S1.2-92-315, ESCA T1S1, contribution.
24. S. Fredricsson, "Beyond FDDI," *Lightwave,* May 1992, pp. 32 ff.
25. F. E. Ross and R. L. Fink, "Overview of FFOL—FDDI Follow-on LAN," *Computer Communications,* Jan/Feb 1992, pp. 5 ff.
26. F. E. Ross and R. L. Fink, "Following the Fiber Distributed Interface," *IEEE Network,* March 1992, pp. 50 ff.

10

Gigabit Systems for Supercomputers: Fibre Channel Standard

10.1 Introduction

Users of supercomputers and large-scale mainframes need to access these systems at gigabit-per-second speeds in order to effectively interact with the data-intensive applications which these supercomputers typically run. Eventually such access will be required at these speeds over WANs, allowing institutions which cannot afford them access on a time-sharing basis. Such WAN connectivity will clearly be based on ATM and SONET. But connectivity starts locally: researchers are now developing premises networks (LANs) to enable workers colocated with these computers unimpeded access. While there is a vendor push to develop premises networks based on ATM principles, as discussed in the previous chapter, some vendors are also pursuing a technology known as fibre channel standard (FCS), particularly in the context of supercomputer applications.

Gigabit networks, including LANs, will allow application designers to treat multiple computing resources as a single system rather than as a network of computers.[1] Some such applications include coupled general circulatory models (to study interaction of atmosphere and oceans), distributed atmospheric models, chemical flowcharting, molecular dynamics, data navigation, calcrust multidatabase integration and imaging, terrain navigation using on-line satellite imagery, collaborative environments and instrument control, radiation treatment planning, radio astronomy, and geographic information systems.[1–5]

The reason for including FCS in this book is that there is a growing need to allow multiple users to reach supercomputers, without requiring these users to be located in the immediate vicinity of the processor.[6,7] In other words, users need high-speed premises distribution networks (i.e., some sort of specialized LAN) to connect to supercomputers. Since the latter are moving away from coaxial-based parallel computer channels to fiber-based serial channels, it is increasingly necessary for workstations to develop interfaces to these supercomputer-specific facilities. Perhaps a LATM network can replace these evolving FCS-based premises networks, or, at least, play a role in this environment. Alternatively, FCS-based premises networks will achieve some penetration of their own. Given this prospect, a primer on FCS is apropos.[8,9]

Initial LATM specifications, discussed in the previous chapter, identified the transmission code used by FCS, known as 8B/10B, as a possible TC sublayer; this approach is known as *block-coded TC*. The rationale for possibly using this approach is to employ already-available FDDI chipsets to bring LATM products to the market quickly. If a LATM vendor were to employ this technology, the LATM network could relatively easily interface to a public B-ISDN using a *physical layer relay* that supported the block-coded TC toward the local network and a SONET TC toward the public network. No rate adaptation is necessary by design. There is no direct technical relationship between LATM and FCS beyond this use of the 8B/10B code.

The X3T9.3 committee of the ANSI is developing a standard for a fiber computer channel, known as FCS. The standard aims at supporting the physical medium levels of a number of existing high-speed interfaces between computers and peripherals. "Fibre" in this context is not the British or international spelling for "fiber," but refers, by definition in the standard, to a *generic* underlying connection mechanism, including, but not limited to, fiber cable. In most cases, the channel will indeed be fiber-based. In the sequel, the term "fiber channel" is used in lieu of the more general "fibre channel"; the terminology we use refers specifically to the channel architecture specified in the FCS.

The channel specified by the FCS provides a transport vehicle for (1) the upper-layer intelligent peripheral interface (IPI) command set, (2) the upper-layer small computer system interface (SCSI) command set, and (3) the high-performance parallel interface (HPPI or HIPPI) data frame. The FCS effort is a response to the need to extend the range of SCSI, IPI, and HPPI, all of which are electrical systems with distance limitations. The FCS is designed to support distances from 2 m to 2 km (optionally 10 km), and to provide switching. The standard, developed in a series of specifications, defines a serial optical and coaxial system operating from 100 Mb/s to 800 Mb/s.

In FCS the channel is optimized for predictable transfers of large blocks of data *over campus-range distances.* This occurs, for example, in the case of file transfer between processors (supercomputers, mainframes, and superminis) and between processors and peripherals. Peripherals include input/output (I/O) storage systems (disk drives and tape units) and high-bandwidth raster scan graphics terminals, output-only devices, such as high-speed laser printers and microfiche devices, and input-only devices, such as optical mass storage devices. Hence, this technology is a mechanism to support high-speed and very high speed interconnection within building, multibuilding, and campus environments. Mechanisms to facilitate "extension" of these channels over metropolitan, regional, or national distances are beyond the scope of the standard itself.

Work on this standard is proceeding at a good pace, with a fairly stable and detailed draft specification already available (revision 2.2, January 24, 1992). Final acceptance and publication of the standard was expected in 1993. Products, such as channel components and complete systems, are already beginning to appear.

10.2 FCS Environment

Figure 10.1 depicts a typical environment to which FCS addresses itself.* In this example, two mainframes located at opposite ends of a large suburban data center need to be connected to support high-speed cooperative computing. For example, the two computers could be located 800 ft away, beyond the reach of traditional bus and tag channels. There may be programmers located on another floor or in another building in a multiacre campus. The programmers need high-throughput access to the mainframes, thereby having to be connected to the mainframe's channel, rather than through a communications controller. There may also be high-end workstations in the network requiring direct channel access.

Three important limitations of existing solutions are providing the impetus for the development of the FCS and for products which implement it. First, there is a multitude of incompatible vendor-specific channels. For example, IBM mainframes have a characteristic channel protocol (at the physical and data link layer level). The IBM channel does not connect with the channel, for example, of a Unisys computer. A Cray computer channel cannot be connected to the channel of a Digital Equipment Corporation supermini. Second, the existing channels, even if they were compatible, have distance limitations, typically in the range of a few hundred feet. This restricts the site or campus topology of the data center and does not allow the remoting of

---

*This discussion is based on Ref. 8.

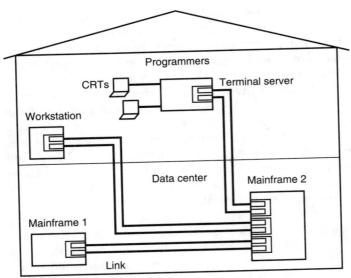

Figure 10.1 Typical environment for use of FCS.

the peripherals over metropolitan or regional areas, as it is often a user's requirement. Third, the existing channels are metallic. These are bulky, expensive, and noise-susceptible cables.

The fiber channel can replace the SCSI, IPI, and HPPI physical interfaces with a protocol-efficient alternative that provides performance improvements in distance and speed. IPI commands, SCSI commands, and HPPI data framing operations can all be intermixed in the fiber channel (proprietary command sets may also share the fiber channel, but this use is not further specified in the FCS). A common compatible lower-level set of protocols is being developed. FCS also specifies the characteristics of the cable, the connectors, the media, the speeds, and other functions required to support reliable communication of the link.

A fiber optic channel medium offers advantages over a metallic medium: high throughput, immunity to crosstalk, low attenuation (implying longer distances), and low error rates. Fiber optic channels replace bulky electrical systems. The fiber cable has a diameter of approximately 1/16 in (complete), while the electrical channel, carrying only a fraction of the bandwidth, has a diameter of 1 in. Several vendors (now including IBM with Enterprise System CONnection, or ESCON) have introduced commercially successful fiber optic channel extenders, thereby proving the concept. The problem has been the utilization of vendor-specific methods and approaches, and the fact that switching can be accomplished only with the addition of expensive external channel matrix switches.

FCS is an important standard to facilitate very high speed connection of processors and peripherals at the campus level. The plethora of vendor-specific protocols currently in place makes generic connection difficult to achieve. High-end users will find that the deployment of channel equipment supporting the FCS will improve their connectivity posture. It is estimated that at the high end, IBM channels represent over 50 percent of the entire market (at the low end, SCSI and IPI markets could conceivably be large, but they are more difficult to size). In this context, it is worth noting that, although IBM has recently announced a vendor-proprietary fiber channel, there is some degree of similarity between it and the FCS, and IBM is an important member of the FCS committee.

FCS hardware modules are beginning to become available from a number of companies. The target cost seems to be $3000 per channel. FCS hardware manufacturers can choose between four sources: (1) short-wave (780 nm) LEDs, (2) long-wave LEDs (1320 nm), (3) short-wave LDs (850 nm), and (4) long-wave (1300 nm) LDs. Each device has its own distance, bandwidth, and cost trade-off. Long-wave lasers allow the highest data rate, but they are expensive ($500 to $1000); long-wave LEDs have a limit of 300 Mb/s, but cost less than $10. The various vendors have opted for different solutions. Several vendors are also implementing prototype switches. These range from systems allowing a single source to transmit at a time, using the entire channel bandwidth, to crossbar switches which have as many as 4096 ports and allow any number of concurrent source-destination paths, each at the full channel bandwidth. HPPI-to-FCS and FCS-to-HPPI adapters are appearing.

Observers are of the opinion that the standard will be implemented relatively quickly by a number of vendors, particularly at the supercomputer and supercomputer workstation level. Introduction of the standard at the mainframe level will take more time because of the entrenched vendor-specific channels; IBM has taken an initial step in that direction with its fiber-based ESCON channel. Use in the SCSI and IPI environments (i.e., minicomputer and microcomputer end) is more difficult to assess.

10.3 Characteristics

10.3.1 Background

The FCS protocol is simple in order to minimize implementation cost and enhance throughput. The transmission medium is isolated from the control protocol so that implementation of point-to-point links, multidrop bus, rings, and cross-point switches may be realized in a technology best suited to the environment of use. The FCS comprises six levels (see Fig. 10.2).

FC-0 defines the physical point-to-point portion of the fiber channel. This includes the fiber, the connectors, and the optical parame-

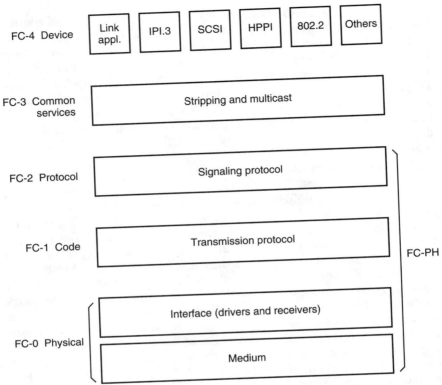

Figure 10.2 Structure of the FCS standards.

ters for a variety of data rates. (A serial coaxial version is also defined for limited distance applications.) The following signaling rates are defined: 132.813 Mbaud, 265.625 Mbaud, 531.25 Mbaud, and 1.0625 Gbaud. These signaling rates correspond to the data transfer rates of 12.5 Mbyte/s (100 Mb/s), 25 Mbyte/s (200 Mb/s), 50 Mbyte/s (400 Mb/s), and 100 Mbyte/s (800 Mb/s). FC-0 operates with a BER of less than 10^{-12}.

FC-1 defines the transmission protocol, which includes serial encoding, decoding, and error control. An 8B/10B scheme is defined, which is similar in concept to the 4B/5B method used in FDDI. Each 8-bit octet is encoded within a 10-bit octet to facilitate error correction and bit synchronization.

FC-2 defines the signaling protocol which includes the frame structure and byte sequences. FC-3 defines the common service interface to FC-4. FC-4 is the highest level in the FCS. It defines the channel protocol, or mapping, between the lower-layer FCS standards and the IPI and SCSI command sets, and the HPPI data framing. FC-F describes the requirements placed on fabrics (generalized switches) which intend to support the fiber channel.

10.3.2 Structure and concepts

The FCS fiber channel is logically a point-to-point serial data channel, designed for high performance. Physically, the channel can support the interconnection of multiple communication points called N_ports (which reside in the end systems), interconnected through a *switching network,* called a fabric. It can also support a simple point-to-point arrangement, without intervening fabric.

The physical interface of the FCS specifies a variety of media and associated drivers and receivers capable of operating at various speeds. The transmission code used is 8B/10B and specified in FC-1. The signaling protocol of the FCS (FC-2) is performed through transfer of frames. This protocol specifies the rules and provides mechanisms needed for end-to-end transfer of block(s) of data. Device protocols constitute FC-4, which is the highest layer in the FCS structure. FC-3 provides services common to multiple device protocols (FC-4s) such as stripping and multicast.

The *fibre channel physical* (FC-PH) layer consists of related functions FC-0, FC-1, and FC-2. Each of these functions is described as a level. As indicated above, "fibre" is a general term used to cover all physical medium types supported by the FCS standard, such as glass fiber, plastic fiber, and copper facilities. The standard does not restrict implementations to specific interfaces between these levels. A node may support one or more N_ports and one or more FC-4s. The FCS provides a method for supporting any number of upper-level protocols (FC-4s).

An N_port is a hardware entity at the node end of the link; it includes a link_control_facility. For example, it could be an IBM mainframe channel or a supercomputer channel. A link consists of two unidirectional fibers transmitting in opposite directions. Such a port may act as an "originator," a "responder," or both, and it contains a transmitter and a receiver. Such a port is assigned a system-unique identifier called N_port identifier, which identifies it during the communication instance. Each N_port contains FC-0, FC-1, and FC-2 functionality. FC-3 optionally provides the common services to multiple N_ports and FC-4s.

A link_control_facility is a link hardware facility which attaches to the end of a link and manages transmission and reception of data. It is contained within each N_port (and F_port, defined next) and includes the link transmission and receiver mechanism.

An F_port is the link_control_facility within the fabric which attaches to an N_port through a link (an F_port is not FC-2 addressable).

A fabric is an entity which allows interconnection of any two N_ports attached to it; that is, it provides switching. The IBM ESCON Director is one example of a recently introduced fabric prod-

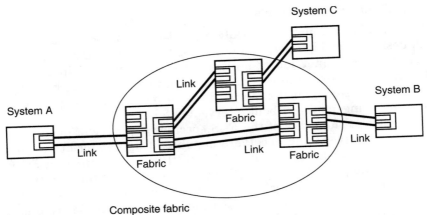

Composite fabric

Figure 10.3 Example of the fabric concept.

uct (although it does not conform completely to FCS). The destination identifier (D_ID) contained in the FC-2 frame header is used to route the data. Figure 10.3 depicts a generic situation with multiple switching elements comprising a "complex" fabric; more typically only one switching element is involved (as shown in Fig. 10.6). When a fabric is present in the configuration, a fiber attaches to an N_port at the user end and an F_port (fabric port) at the fabric end. An end-to-end communicating route may be made up of physical links of different technologies. For example, it may have multimode fiber links attached to end ports but may have a single-mode link in between. In the sections which follow, some highlights from the key FC-0, FC-1, and FC-2 standards are provided.

10.3.3 FC-0 general description

The FC-0 level of the standard describes the physical link of the fiber channel. The FC-0 covers a variety of media, and the associated drivers and receivers capable of operating at a wide range of speeds. The FC-0 is designed for maximum flexibility and allows the use of a number of technologies to meet a wide range of system requirements. Each fiber is attached to a transmitter of a port at one end and a receiver of another port at the other end, as seen in Fig. 10.4.

The function of an FC-0 transmitter is to accept a parallel input from a connected device, along with the clock, and convert the signal to a serial stream. The signal is then modulated over the appropriate underlying medium (typically a fiber). Three interfaces associated with the transmitter are the parallel input, the serial output, and the medium output, following modulation.

The function of an FC-0 receiver is to demodulate the signal from

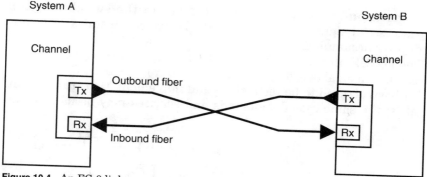

Figure 10.4 An FC-0 link.

the appropriate medium (typically a fiber). The clock embedded in the signal by the transmitter must be recovered. The serial bit stream must then be converted to parallel, which is passed along, with the clock, to the destination device. Four interfaces associated with a receiver are medium input, receiver output, clock recovery output, and parallel output.

Transmission characteristics

Single-mode cable plant usage. The FC-0 data link can support two distances. A FC-0 link with a nominal distance capability of 10 km has 14 dB loss budget; data links with a nominal distance of 2 km have 6 dB loss budgets. The cable has a loss of 0.5 dB/km; the connectors also create a loss of 0.6 dB per end. Splices also can add to the attenuation (typically 0.4 dB per splice).

Multimode cable plant usage. This configuration supports a maximum optical power budget of 12 dB for laser diode (LD) links and 6 dB for light-emitting diodes (LED) links. Four cable versions are supported, two with LEDs and two with LDs as follows:

1. 50/125 μm, emission at 780 nm, loss of 4.0 dB/km
2. 62.5/125 μm (same as FDDI), emission at 780 nm, loss of 4.5 dB/km
3. 50/125 μm, emission at 1300 nm, loss of 1.5 dB/km
4. 62.5/125 μm, emission at 1300 nm, loss of 1.5 dB/km

In addition to these cable losses, there are connector and splice losses. These parameters result into links which range from 350 m to 2 km.

FC-0 also specifies the data rates of the serial data stream, since it must accept and recover the clock. The rates, already identified, are 132.813 Mbaud, 265.625 Mbaud, 531.25 Mbaud, and 1.0625 Gbaud.

An elastic buffer is required at all nodes in the switch fabric or repeaters. All signals must be retimed before retransmission to prevent jitter accumulation.

The logical interface of FC-0 to FC-1 occurs at the serial data interfaces, described above. The interface between FC-0 and FC-1 is intentionally structured to be technology and implementation transparent. That is, the same set of commands and services may be used for all

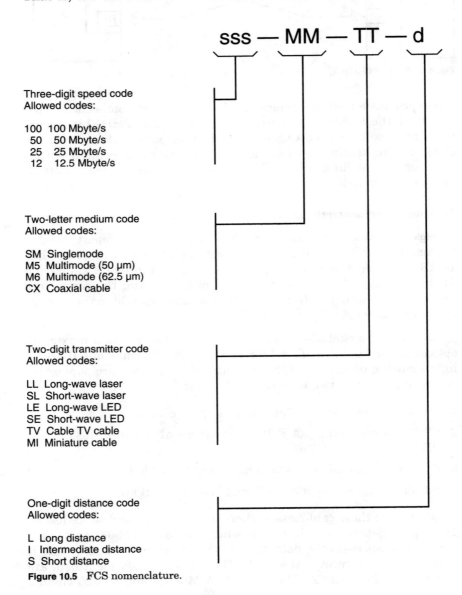

Three-digit speed code
Allowed codes:

100 100 Mbyte/s
 50   50 Mbyte/s
 25   25 Mbyte/s
 12   12.5 Mbyte/s

Two-letter medium code
Allowed codes:

SM Singlemode
M5 Multimode (50 µm)
M6 Multimode (62.5 µm)
CX Coaxial cable

Two-digit transmitter code
Allowed codes:

LL Long-wave laser
SL Short-wave laser
LE Long-wave LED
SE Short-wave LED
TV Cable TV cable
MI Miniature cable

One-digit distance code
Allowed codes:

L Long distance
I Intermediate distance
S Short distance

Figure 10.5 FCS nomenclature.

signal sources and communications media. This allows the interface hardware to be interchangeable at the system level without having to be concerned with the technology of a particular implementation.

The nomenclature for the FC-0 technology options described in the FCS is shown in Fig. 10.5. *Long distance* refers to the range of 2 m to 10 km; *intermediate distance* refers to the range to 2 m to 2 km; *short distance* refers to the range 2 m to 50 m. Table 10.1 depicts presently supported signal interface types for both fiber and coaxial media. Table 10.2 depicts currently supported communication media. As can be seen from these tables, FC-0 provides for a large variety of distances, cable plants, and technologies.

10.3.4 FC-1 general description

Two sublayers (function groupings) of the physical layer are PMD and TC. As described earlier, TC deals with aspects of this layer that are independent of the actual characteristics of the medium (these being handled by PMD).

The FCS ensures that the transmission characteristics of information to be transmitted over a fiber are sufficiently robust through the use of a transmission code. This transmission code accepts unencoded data and converts it to a form which ensures that sufficient transitions are present on the encoded bit stream to allow clock recovery at the receiver. Certain encodings specified by the transmission code have special characteristics which allow a receiver to easily determine word alignment on the incoming bit stream.

Transmitter and receiver behavior is specified via a set of states and their interrelationships. These states are divided into operational and not-operational classes. Error monitoring capabilities and special operational modes are also defined for operational receivers and transmitters.

TABLE 10.1 FCS Interfaces Supported

Fiber systems	Coaxial systems
100-SM-LL-L	100-CX-TV-S
100-SM-LL-I	50-CX-TV-S
50-SM-LL-L	25-CX-TV-S
50-M5-SL-I	12-CX-TV-S
25-SM-LL-L	
25-SM-LL-I	
25-M6-LE-I	
25-M5-SL-I	
12-M6-LE-I	

TABLE 10.2   FCS Cable Plants Supported

Single-mode fiber	Multimode (50 μm)
100-SM-LL-L	50-M5-SL-I
100-SM-LL-I	25-M5-SL-I
50-SM-LL-L	25-M5-LE-I
25-SM-LL-L	12-M5-LE-I
25-SM-LL-I	
Multimode (62.5 μm)	**Coaxial**
50-M6-SL-I	100-CX-TV-S
25-M6-SL-I	50-CX-TV-S
25-M6-LE-I	25-CX-TV-S
12-M6-LE-I	12-CX-TV-S
	100-CX-MI-S
	50-CX-MI-S
	25-CX-MI-S
	12-CX-MI-S

Information to be transmitted over a link is encoded 8 bits at a time into a 10-bit transmission character and then sent serially. Information received over a link is collected 10 bits at a time, and those transmission characters that are used for data, called data characters, are decoded into the correct 8-bit codes. The 10-bit transmission code provides for all two hundred fifty-six 8-bit word combinations. Some of the remaining transmission characters, referred to as special characters, are used for functions which are to be distinguishable from the contents of a frame. This encoding scheme is called 8B/10B nonreturn-to-zero code and is identical to that employed in the IBM's ESCON environment. Consider, as an example, the character "E." The EBCDIC encoding of "E" is Hex C5, i.e., 11000101 (for bits 0, 1, 2, ..., 7, respectively). Converted to 8B/10B notation this character becomes:

Step 1.   Reverse the bits: 10100011.

Step 2.   Partition the bits into two sets: 10100 and 011.

Step 3.   Concatenate 1 to the tail of the first set a 0 at the tail of the second set, i.e., 101001 and 0110.

Step 4.   Recombine to obtain the final code 1010010110.

10.3.5   FC-2 general description

The fiber channel is logically a point-to-point serial data channel, structured for high-performance capability. Physically, it can be a

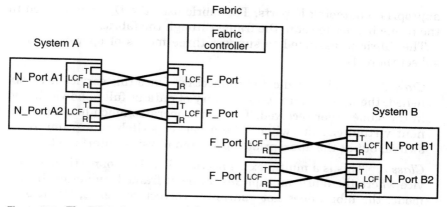

Figure 10.6 The FC-2 physical model. LCF = link_control_facility; T = transmitter; R = receiver.

point-to-point system or it could be composed of a set of links through a switching fabric, as already shown in Fig. 10.3. The FC-2 level serves as the transport mechanism of the fiber channel. The transported data is transparent to FC-2 and visible to FC-3 and above.

Figure 10.6 depicts the FC-2 physical model. The fiber channel physically consists of a minimum of two nodes (say a mainframe and a peripheral), each with a minimum of one N_port interconnected by a pair of fibers (one outbound and the other inbound at each N_port) constituting the link. The link is used by the interconnected N_ports to perform data communication.

User equipment such as a computer, a controller, a peripheral, or a high-speed terminal can be interconnected to other equipment through these links. Each node contains one or more N_ports (in the example, each node has two N_ports). The FC-2 supports simultaneous, symmetrical bidirectional flow. The fabric allows switching. An N_port logically performs only point-to-point communication with another N_port at any instant in time, even in the presence of a fabric. However, multiple N_ports in a node can simultaneously perform data transfers with a single or multiple N_ports contained in one or more nodes. This allows the user equipment to perform simultaneous parallel data transfers in a flexible manner to or from a single or multiple attached devices.

Each link_control_facility must be able to perform the logical and physical control of the link for each mode of use and provides the logical interface to the rest of the end system. It is basically a termination card with hardware, software, and firmware to support the FCS protocols.

Data is transmitted in frames. The N_port sends frames and link control responses to frames it receives. The point-to-point topology (without intervening fabric) is clearly the simplest situation. In the presence of a fabric, the fabric must be able to route frames to the appropriate outgoing F_ports. The fabric uses the D_ID embedded in the frame header to route the frame through the fabric.

The fabric is assumed to support three modes of operation (or a subset thereof):

Class 1. This is a dedicated connection. Once this service is established, the fabric guarantees delivery of data at full bandwidth and in the same order received. The fabric operates much like a permanent virtual circuit switch (like a packet switch or a frame relay switch, but at the appropriate speed and physical interfaces).

Class 2. This is a multiplex connection. This is a connectionless service where the fabric multiplexes frame at frame boundaries. In this mode, the fabric does not guarantee correct sequence, but it does guarantee notification of delivery or failure to deliver, in the absence of link errors (in case of link errors, notifications are not guaranteed).

Class 3. This is a connectionless service where the fabric multiplexes frame at frame boundaries. It supports open-ended delivery where the destination N_port does not send any confirmation link control frames.

The primary function of the fabric is to receive frames from a source N_port and to route the frames to the destination N_port whose address is specified in the frame. The fabric may or may not verify the validity of the frames as they pass through the fabric. FC-2 specifies the protocol between the fabric and the attached N_ports. The F_port is the fabric's mechanism for physically connecting the user's N_port. The receiving F_port responds to the sending N_port according to the FC-2 protocol. An F_port may or may not contain a receiver buffer for the incoming frames.

A set of building blocks, their interrelationships, behavior rules, and usage hierarchy are defined in FC-2. These building blocks are frame, sequence, exchange, and protocol.

Frame. Frames are based on a common format. Frames can be categorized in the following classes: data frames and link control frames. Link control frames are further classified as acknowledge frames and link response frames (selective retransmission of frames is not supported by FC-2).

Sequence. A sequence is a set of one or more related data frames transmitted unidirectionally from one N_port to another N_port, with corresponding link control frames, if applicable, transmitted in response. Error recovery is performed at the sequence boundary. If a frame is received with an error, and if the error policy requires error recovery, the sequence to which the frame belongs is retransmitted. Sequences are identified by a sequence identifier (SEQ_ID).

Exchange. Exchanges are composed of one or more nonconcurrent sequences. An exchange is the basic mechanism used to transfer data between two N_ports. An exchange may be unidirectional or bidirectional.

Protocol. FC-2 provides data transfer protocols to be used by higher-level protocols to accomplish a given level of function. The standard also provides log-in and log-out control to manage the operating environment to perform data transfers.

Frame structure. An ordered set is a 4-byte combination of data and special transmission characters which is designated by the standards to have special meaning. Ordered sets provide the ability to establish word-boundary alignment. The actual encoding of characters is described in FC-1. Ordered sets can be used to define the following control "characters" which are required for the data link protocol:

1. Start-of-frame delimiters
2. End-of-frame delimiters
3. Primitive signals
   - Idle
   - Receiver ready
4. Primitive sequences
   - Not operational
   - Offline
   - Link reset
   - Link reset response

A frame delimiter is an ordered set that immediately precedes or follows the contents of a frame. Separate and distinct delimiters identify the start of a frame and the end of a frame. The start-of-frame (SOF) delimiter is an ordered set that immediately precedes the frame content. The end-of-frame (EOF) delimiter is an ordered set that immediately follows the frame trailer. The EOF delimiter designates the end of the frame content and is followed by idle words.

All FC-2 frames follow the general frame format as shown in Fig. 10.7. An FC-2 frame is composed of a start-of-frame delimiter, frame content, and an end-of-frame delimiter. The frame content is composed of a frame header, a data field, and a 32-bit CRC.

Figure 10.7   FC-2 frame.

10.4   Conclusion

Rapid progress is being made in standardizing a high-speed channel for efficient support of direct local connectivity between computers and between computers and peripherals. The increased throughput requirement typical of evolving high-end applications, such as computer-aided design, multimedia, animation, and visualization, make this standard an important one.

References

1. C. E. Catlett, "In Search of Gigabit Applications," *IEEE Communications Magazine,* April 1992, pp. 42 ff.
2. J. B. Lyles and D. C. Swinehart, "The Emerging Gigabit Environment and the Role of Local ATM," *IEEE Communications Magazine,* April 1992, pp. 52 ff.
3. N. K. Cheung, "The Infrastructure for Gigabit Computer Networks," *IEEE Communications Magazine,* April 1992, pp. 60 ff.
4. H. T. Kung, "Gigabit Local Area Networks: A Systems Perspective," *IEEE Communications Magazine,* April 1992, pp. 79 ff.
5. P. Newman, "ATM Technology for Corporate Network," *IEEE Communications Magazine,* April 1992, pp. 90 ff.
6. D. Getchell and P. Rupert, "Fiber Channel Local Area Network," *IEEE LS,* May 1992, pp. 38 ff.
7. M. Fahey, "Suppliers Focus on Fiber Channel," *Lightwave,* September 1992, pp. 42 ff.
8. D. Minoli, "ANSI Fibre Channel Standard," Datapro Communications Series: Broadband Networking, Report #2070, 4/1992.
9. D. Minoli, "IBM Enterprise Systems Connection (ESCON)," Datapro Communications Series: Broadband Networking, Report #2080, 4/1992.

Index

ABOUT THE AUTHOR

Daniel Minoli has spent the past eight years at Bell
Communications Research (Bellcore) as a strategic data
communications planner. His research has been aimed at
supporting the internal data processing/data communications needs of the Bell Operating Companies; identifying
data services that can be provided using ISDN, B-ISDN,
AIN, and other platform-independent infrastructures; network design tasks for large end-user networks; and
B-ISDN/ATM signaling standards work. In addition, he has
worked for Bell Telephone Laboratories, ITT World
Communications, and Prudential-Bache Securities. He is
the author of several communications books and a con-
tributing editor to *Network Computing* magazine.